(*continued on back*)

Two-Dimensional NMR Spectroscopy

CHEMICAL ANALYSIS

A SERIES OF MONOGRAPHS ON
ANALYTICAL CHEMISTRY AND ITS APPLICATIONS

VOLUME 97

A WILEY-INTERSCIENCE PUBLICATION

JOHN WILEY & SONS

New York / Chichester / Brisbane / Toronto / Singapore

Two-Dimensional NMR Spectroscopy

JAN SCHRAML

Institute of Chemical Process Fundamentals
Czechoslovak Academy of Sciences
Prague, Czechoslovakia

JON M. BELLAMA

Department of Chemistry and Biochemistry
University of Maryland
College Park, Maryland

WILEY

A WILEY-INTERSCIENCE PUBLICATION

JOHN WILEY & SONS

New York / Chichester / Brisbane / Toronto / Singapore

Library of Congress Cataloging in Publication Data:
Schraml. Jan.
    Two-dimensional NMR spectroscopy / Jan Schraml. Jon M. Bellama.
        p.   cm.
    "A Wiley Interscience publication."
    Bibliography: p.
    Includes index.
    ISBN 0-471-60178-0
    1. Nuclear magnetic resonance spectroscopy.   I. Bellama, Jon M.
    II. Title.
    QD96.N8S37 1988
543′.0877—dc19   87-33315
                                                    CIP

PREFACE

This book originated from various lectures and discussions that the authors held on several occasions. Most notable were the 2D NMR Summer School organized by the Czechoslovak Spectroscopic Society at Kočovce (1983) and the courses on 2D NMR spectroscopy given at the Fachhochschule Münster, which provided valuable feedback from the students and faculty. J.S. is grateful to the Fachbereich Chemieingenieurwesen for several awards of visiting professorships (1983–1986), during which most of his part of the book was written.

Some parts of this book were originally published in a Czech version. The authors wish to thank Academia, the publishing house of the Czechoslovak Academy of Sciences, for their kind permission to use that material in the present text.

The authors also appreciate very much the assistance of the scientific exchange program sponsored by the National Academy of Sciences (USA) and the Czechoslovak Academy of Sciences, and a travel grant awarded by the University of Maryland to J.M.B., all of which made possible the collaboration of the authors.

We are also indebted to a number of individuals. J. Keeler and G. A. Morris kindly provided programs, and we appreciate discussions with M. H. Levitt and G. Bodenhausen. R. Freeman, H. Kessler, G. A. Gray, and M. A. Bernstein provided us preprints of work in press, and the 500-MHz spectra were run by T. Pehk. Finally, we wish to thank P. Sedmcra, M. Holík, M. Buděšínský, and V. Blechta for discussions, constructive criticism of the manuscript, and help in running the demonstration spectra that appear throughout the book.

JAN SCHRAML
JON M. BELLAMA

Prague, Czechoslovakia
College Park, Maryland

v

CONTENTS

Two-Dimensional NMR Spectroscopy

CHAPTER

1

INTRODUCTION

The new and very powerful technique of two-dimensional nuclear magnetic resonance (2D NMR) spectroscopy is one of the major advances in the field of NMR spectroscopy in recent years. 2D NMR spectra are now found routinely in chemical journals. One example of a 2D spectrum that might be found in a chemistry journal is the interesting picture shown in Fig. 1.1.

The publishing of such spectra, of course, consumes precious journal space, and editors like to hold the number of published figures to a minimum. 2D NMR is an exciting new technique, however, that brings information usually not accessible by other means. Unless you are an expert or at least are familiar with 2D NMR spectroscopy, however, the legends that describe the 2D NMR spectra may not help you very much in appreciating the information that the 2D NMR spectrum contains. The legend may contain acronyms (e.g., COSY, NOESY, SECSY, DQCOSY, INADEQUATE, etc.) and other specialized terms that may not make much sense.

The goal of this book is to explain the fundamentals of two-dimensional NMR spectroscopy to the extent that you should be able to "read" 2D NMR spectra (i.e., to obtain the information that is encoded in the 2D NMR spectrum) and to have some understanding of the factors that cause a 2D NMR spectrum to have the appearance that it has. Also, after reading this book you should be able to decide which of the many measuring techniques of 2D NMR spectroscopy will help solve your problem and to use manufacturer-provided software to run such an experiment on a commerical spectrometer.

To achieve this goal we have sacrificed mathematical rigor and physical exactness, which demand that 2D NMR spectroscopy be treated in terms of a density matrix. There are books already available for a mathematically or physically oriented person who is well versed in density matrix or product operators. The book by Bax [1], for example, is well known, and a rigorous treatment can also be found in the book by Ernst et al. [2].

To keep the present book to a reasonable size, we must assume a certain working knowledge of conventional one-dimensional (1D) NMR spectroscopy. Presumably, you are already familiar with the concepts of

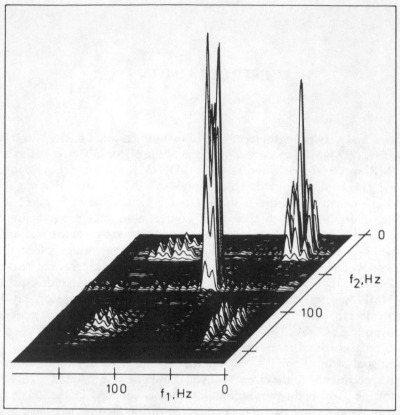

Figure 1.1 COSY-90 ^{1}H NMR spectrum of 2-butanol (portion of a whitewashed stacked-trace plot of a power spectrum).

"chemical shift" and "coupling constants." If some terms related to NMR experiments do not appear to be sufficiently clear, you should brush up on these terms by consulting the Appendix or other books such as refs. 3 to 5.

2D NMR represents a most significant achievement in the continuing effort to increase spectral resolution in NMR spectroscopy. High spectral resolution, that is, narrow lines dispersed over a large frequency span (spectral width in hertz), is needed for analytical applications and for facile interpretation of spectra. Since the "invention" of NMR in 1946, the spectral resolution of NMR spectrometers has been increased continuously. The successful efforts to improve the homogeneity and stability of the magnetic field and temperature over the entire sample have resulted in very narrow NMR lines (with the ultimate apparently achieved recently by Allerhand et

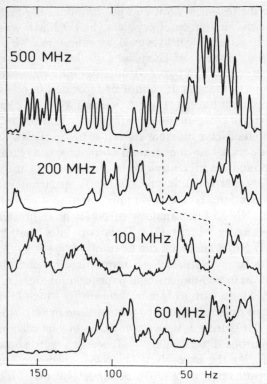

Figure 1.2 Portions of one-dimensional ^1H NMR spectra of an unknown trisaccharide measured on spectrometers with different operating frequencies: 60 MHz, Tesla BS-467; 100 MHz, Tesla BS-497; 200 MHz, Varian XL-200; 500 MHz, Bruker AM-500. Sections to the right of the dashed line correspond to the entire 500-MHz spectrum as shown

al. [6], who reported contributions to the linewidth from instrumental factors as small as 7 mHz). The continuing efforts to increase the intensity of the magnetic field (the higher the magnetic field induction, the larger the spectral width) have produced commercial spectrometers that operate at 600 MHz for proton NMR measurements. The effect of increasing magnetic field on the spectral resolution and its importance for the interpretation of spectra are demonstrated in Fig. 1.2. Although considerable improvement in spectral resolution is noted as the operating frequency increases, the lines are still not resolved in the 500-MHz spectrum, and further efforts in this direction are not likely to bring about any dramatic improvements. A twofold increase in spectral width (in hertz) would require a twofold increase in magnetic field (and even this twofold increase would not be sufficient to resolve the lines in the region 0 to 50 Hz in Fig. 1.2). Furthermore, such an increase would not only be close to the technical limit and not economically feasible [7], but it would not solve the fundamental problem. In 1D NMR spectroscopy, no matter how strong a magnetic field is used, we cannot resolve two singlet lines that accidentally have the same chemical shift.

The analogous situation in paper chromatography was solved many years ago. If one group of compounds cannot be separated in one solvent system because the compounds have the same R_F values (R_{F1}), the paper is turned by 90° and the individual compounds are separated in another solvent system in which the compounds have different R_F values (R_{F2}). This chromatographic experiment produces a two-dimensional chromatogram with the spot of a particular compound described by two coordinates, R_{F1} and R_{F2}.

A similar effect, the spreading of the NMR spectrum in a second and perpendicular direction, is the technique used to increase the spectral resolution in 2D NMR. (The analogy of NMR and chromatography was originally proposed by Morris [8].) Thus two lines that have the same frequency in a 1D NMR spectrum can be resolved by spreading the spectrum into another dimension. Naturally, accomplishing this effect in NMR is not as easily achieved as in two-dimensional paper chromatography. Most of this book is concerned with various technical means by which NMR spectra can be spread into two dimensions; at the moment, we merely note how efficient these means are. Figure 1.3 shows the 2D NMR spectrum of the same sample for which one-dimensional NMR spectra were shown in Fig. 1.2. Obviously, spreading the unresolved 1D NMR spectrum measured at 200 MHz in another direction gave much better resolution of individual peaks than that achieved even at 500 MHz in a one-dimensional spectrum. The lines in the region 0 to 50 Hz, which were not resolved at 500 MHz in the 1D spectrum, are well resolved in the 2D spectrum.

It seems to be agreed that the idea of spreading NMR spectra into two dimensions was presented for the first time by Jeener [9], who suggested the method in 1971 as a convenient alternative to a series of decoupling experiments. Jeener's idea was to obtain a two-dimensional spectrum by a two-dimensional Fourier transform of a signal recorded as a function of two time variables. A detailed analysis by Ernst's group [10] has shown that the original idea of Jeener has indeed a much broader applicability in NMR spectroscopy, and it can be applied in other fields of spectroscopy as well. In NMR spectroscopy it can be used to increase the spectral resolution to reveal in a very straightforward way NMR parameters that are important for structure elucidation, to determine quantities that cannot be observed directly, and many other applications that will be discussed in subsequent chapters. The methods of 2D NMR are also well suited for automation; the experiments can be run unattended by a spectrometer operator.

At present, there are scores of 2D NMR methods available, and new ones

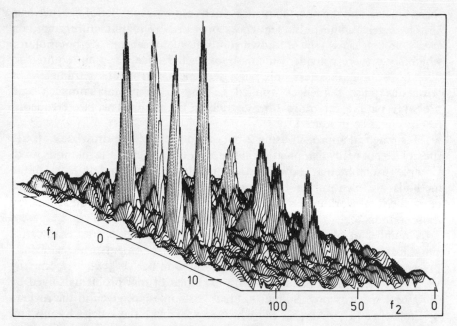

Figure 1.3 Portion of a J-resolved 2D NMR spectrum of an unknown trisaccharide; ^{1}H NMR, 200 MHz, the same sample as in Fig. 1.2, all values in hertz.

appear frequently. All of them are based on the original idea of Jeener, but they use different means (pulse sequences) to record the NMR signal as a function of two time variables. The pulse sequence used in an experiment determines the physical meaning of the two coordinates in the 2D spectrum, or, in other words, the type of information contained in the 2D spectrum. We shall limit ourselves to those types of 2D NMR spectroscopy that apply to magnetic resonance of nuclei with spin $\frac{1}{2}$ in liquid isotropic samples (i.e., we shall not be concerned with those forms of 2D NMR spectroscopy that are specific for nuclei with quadrupole moments, or for measurements in solids or of oriented molecules).

We demonstrate the various methods of 2D NMR spectroscopy by using only one simple example, a concentrated solution of 2-butanol (*N,N*-dimethylacetamide is also used in Section 4.3). Since this is a readily available compound, you should be able to reproduce our results on your own spectrometer (for sample preparation, see Section A.9). This is not as simple as it might first appear. A certain drawback of 2D NMR spectroscopy is that the measurement and plotting of spectra require a considerable number of parameters to be specified to the computer that controls the spectrometer, and the pulse sequences each contain a number of instructions. Unfortunately, some pulse sequences provided by manufacturers contain errors, and measurement of known solutions of concentrated 2-butanol, for which spectra are provided in this book, will reduce the time required to "debug" your parameters and pulse sequences. Erroneous parameters or erroneous pulse sequences applied to any experimental sample would probably require far more time to debug, or, perhaps worse, erroneous results could be produced.

This practical approach using 2-butanol does have one drawback; its 2D spectra do not really demonstrate sufficiently the power of the methods used. Complex examples that show more adequately the power and potential of the methods are plentiful in the literature, but you would not be able to reproduce that work unless you synthesize the particular compound used by those authors (e.g., a tricyclodecane derivative [1]).

A number of other works on 2D NMR are available. For example, a simplified approach to the physics of 2D NMR experiments was beautifully presented in the review article by Benn and Günther [11], who accompany very clear pictorial explanations with examples of real problems solved by 2D NMR spectroscopy. Similarly, "real" examples are given in the reviews by Freeman and Morris [12–14], which together with the articles by Bax [15] and Turner [16] can help the reader to bridge the gap between a simplified treatment and the theoretical approach presented in refs. 1 and 2. The review by Morris [14] compares 2D NMR with 1D NMR techniques for structure elucidation. Pulse sequences used in NMR spectroscopy of liquids have been reviewed by Turner [17]. Two collective volumes of treatises [18, 19] contain well-written 2D NMR introductory chapters [20, 21] sections on experimental aspects [22], and a description of strategy for any 2D NMR technique [23], as well as chapters on applications to various fields of chemistry (e.g., saccharides) [24]. Other reviews that cover applications of 2D NMR to specialized branches of chemistry such as stereochemistry [25] and peptides [26] are also available.

CHAPTER

2

FUNDAMENTALS OF 2D NMR SPECTROSCOPY

2.1. 2D NMR AS AN EXTENTION OF CONVENTIONAL PULSED 1D NMR SPECTROSCOPY

If we compare the procedures used to obtain 1D NMR and 2D NMR spectra, it is astonishing that the idea of 2D NMR did not occur many years earlier; 2D NMR now appears to be quite straightforward. The procedure used to measure a conventional 1D NMR spectrum on a pulsed spectrometer is shown schematically in Fig. 2.1. Some details of this procedure are reviewed in Appendix Section A.1.)

During the preparation period a thermal equilibrium is established among the nuclear spins in the sample. The spin system is then disturbed by a radio-frequency (rf) pulse. After the pulse we follow the motion of the spin system during the detection period. We can monitor the spin system as its motion induces a signal that is detected in the receiver of our NMR spectrometer. The time dependence of the signal obtained, $s(t)$, is known as an FID (free induction decay), and a Fourier transform (FT) of the FID produces a frequency dependence of the signal, $S(f)$ (i.e., the usual 1D NMR spectrum).

Since our desired two-dimensional spectrum is a signal dependence on two independent frequencies, $S(f_1, f_2)$, it is obvious that it could be obtained by a Fourier transform of the signal dependence on two independent time variables, t_1 and t_2. The problem is how to measure the dependence of the NMR signal on two time variables, $s(t_1, t_2)$. Let us pursue the following oversimplified line of thought: after one pulse, we get a one-dimensional dependence of the signal on time; hence, by simple extrapolation we can expect to obtain a two-dimensional dependence if we use two pulses. This is exactly the experiment proposed by Jeener [9]. This experiment, or more correctly the principle involved in this experiment, has become the basis of 2D NMR spectroscopy.

The Jeener experiment is shown in Fig. 2.2. As in 1D measurements, the experiment starts with a preparation period to establish thermal equilibrium. The first pulse disturbs the spin system. After the pulse, the spin system evolves in the same way that it does in a 1D measurement, but we do not

7

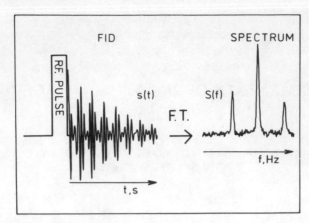

Figure 2.1 Pulsed 1D NMR measurement. The pulse is preceded by a preparation period; time is measured in seconds, s; the signal intensity is in arbitrary units (e.g., in volts, V); FT stands for Fourier transform; the unit of frequency f is hertz, Hz)

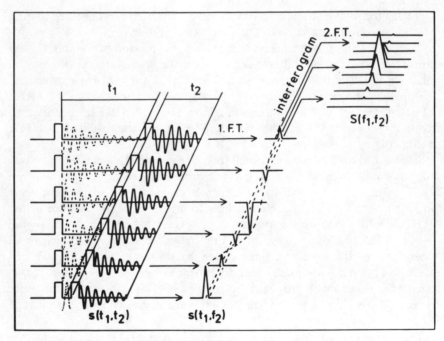

Figure 2.2 The Jeener experiment.

record or directly follow this evolution. In contrast to a 1D experiment, however, the evolution of the spin system is interrupted by a second rf pulse at time t_1 (measured from the end of the first pulse). After the second pulse, the spin system again evolves, and we detect and record the signal during time t_2 (measured from the end of the second pulse) of the detection period. The FID detected [i.e., the dependence of the signal on time t_2, $s(t_2)$] also depends on the instant at which the evolution of the spin system has been interrupted in the evolution period (i.e., on the duration of evolution period t_1).

For example, if the second pulse occurs immediately after the first one is completed ($t_1 = 0$), the FID corresponds to a one-dimensional measurement with a doubled pulse length. It should also be apparent that if time t_1 is so long that the spin system has returned to equilibrium before the second pulse is applied, the FID detected is the same FID that would be obtained in a one-dimensional measurement with a single pulse of the same pulse length.

Repeating this experiment for a set of different evolution times t_1, we measure the dependence of the signal on the two time variables, $s(t_1, t_2)$. A Fourier transformation (one-dimensional) of every FID with respect to t_2 yields a set of spectra, each of which corresponds to a particular evolution time t_1. Mathematically, the FT converts functions $s(t_2)$ into functions $S(f_2)$, each with a different t_1 parameter. The whole set represents a function of the signal with respect to time and frequency, $s(t_1, f_2)$. On a graph the set of spectra looks very much like a 2D spectrum (see Fig. 2–3), but it isn't. It is only a series of spectra that have been measured for different values of evolution time; signals in these spectra vary with time t_1. In the example shown in Fig. 2.3, the intensities of the signals vary (amplitude modulation) with the evolution time in the same way that the intensity of a FID varies with the detection time (damped oscillations).

In order to see the similarity of the spectrum and the FID, it is necessary to compare signals with the same frequency f_2 in different spectra measured for different evolution times, t_1. This rearrangement of the data (transposition) corresponds to viewing the data shown in Figs. 2.2 and 2.3 in a skewed direction along the t_1 axis. To distinguish the dependence of the signal on time parameter t_1 from the FID, which the signal resembles, the dependence is often called an interferogram in the literature on 2D NMR spectroscopy. A Fourier transform (the second FT) of all interferograms produces the desired two-dimensional spectrum [i.e., the function $S(f_1, f_2)$].

Our simplified approach has led to the correct result only because the spin system has a certain memory. The behavior of the spin system during the detection period t_2 is affected by the past history of the system, in particular by the events that have taken place during the evolution time t_1.

The first rf pulse induces a certain coherence among nuclear magnetic

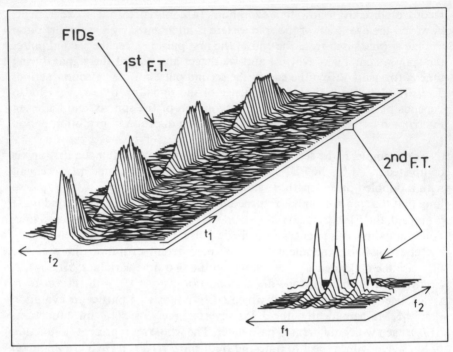

Figure 2.3 Example of the spectral dependence, $s(t_1, f_2)$, on evolution time t_1. The evolution time has been incremented by 3 ms for each consecutive experiment; amplitude modulation, absolute value.

moments. After the pulse, these magnetic moments do not have the random orientations that they had before the pulse. Instead, they have fixed phases (see Section A.1). The same phase relationship among the moments (coherence) is maintained for a considerable length of time; during this time the motion of the spins is coherent. The coherence is gradually lost by spin-spin relaxation processes with a characteristic spin-spin relaxation time T_2. (Do not mistake T_2 for t_2, the detection time!) The persistence of the memory of the spins for some time (the memory of the spin system) is the reason that the motion of the spins during the detection period is affected by their motion in the preceeding evolution period. Each spectral component, which precesses with a certain characteristic precession frequency during t_1, passes on the information about its motion or evolution during this time to some appropriate spectral component that we measure during t_2. Thus information about motion during t_1, when no signal is being recorded, is obtained indirectly through the effect that the t_1 time has on the signal measured during t_2. Freeman [12, 13] compares this process of indirectly obtaining the

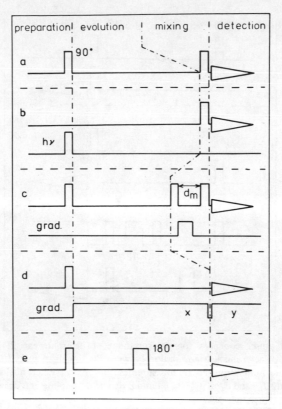

Figure 2.4 Some pulse sequences for measurement of homonuclear 2D NMR spectra: (*a*) Jeener experiment (COSY); (*b*) measurement of chemical reaction (CIDNP); (*c*) exchange 2D spectroscopy with a magnetic field gradient pulse; (*d*) "imaging," a measurement of spin density in a plane using two magnetic field gradients; (*e*) measurement of J-resolved spectra. Note that pulse sequences (*d*) and (*e*) do not contain a mixing period; single-width rectangles denote 90° pulses, and double-width rectangles indicate 180° pulses.

information to a relay race where each runner passes a baton to the teammate who is running the next stage of the race.

A detailed analysis of the Jeener experiment was carried out by Ernst's group [10] in a fundamental paper that is now recognized as a classic in the field of 2D NMR spectroscopy. We consider the specific conclusions regarding the Jeener experiment later (Section 2.4), but first we address 2D NMR spectroscopy in general.

To measure the signal dependence on two time variables t_1 and t_2, $s(t_1, t_2)$, it is necessary that some physical events mark two time intervals t_1 and t_2 on the time axis. In general, the time axis can thus be broken into four intervals (Fig. 2.4) that are known as the preparation, evolution, mixing, and

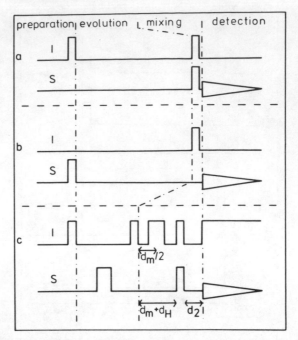

Figure 2.5 Some pulse sequences for measurement of heteronuclear 2D NMR spectra: (a) measurement of heteronuclear correlation between the signals of spins I and S; (b) a two-pulse heteronuclear pulse sequence, (c) measurement of RELAY heteronuclear spectra. $d_m = 1/(2 \cdot J_{II})$, $d_H = 1/(2 \cdot J_{IS})$ and $d_2 = 1/(3 \cdot J_{IS})$; I and S denote decoupling and observation channels of the spectrometer, respectively.

detection periods, which usually occur consecutively during the course of a 2D NMR experiment.

In a 2D NMR experiment the length of evolution period t_1 is varied in a systematic manner. For each value of t_1, the signal dependence on time t_2 is measured (acquired) during the detection period. The entire set of these dependences for all values of t_1 represents the function $s(t_1, t_2)$. A Fourier transformation of this function with respect to both t_1 and t_2 produces a two-dimensional spectrum, $S(f_1, f_2)$. This is the general procedure for measuring 2D spectra.*

Different physical events can be used to mark the beginning or the end of a

---

*Although it is confusing, it is common practice to regard t_1 and t_2 not only as running variables of the two time periods but also as their maximum values. We have accepted this rather misleading practice to help the reader understand the literature of the field. When it is necessary to distinguish the maximum values, they are denoted by the superscript "max."

particular time period. A change of magnetic field, a rf pulse, the switching on or off of some spectrometer function (e.g., receiver or decoupler), a pulse of light, or other physical events are all possible.

The entire set of conditions and physical events to which the spin system is subjected during the experiment is termed the "pulse sequence." Some pulse sequences used in 2D NMR spectroscopy are shown in Figs. 2.4 and 2.5. The examples of pulse sequences are divided into two groups. In Fig. 2.4 are examples of homonuclear measurements in which we follow the behavior of nuclear species of the same kind (most frequently protons) during both t_1 and t_2 intervals. In heteronuclear experiments (Fig. 2.5) the pulses affect nuclei of different types.[†] Usually, one type of nuclei is followed during t_2 (S nuclei), and the other is either followed indirectly through time t_1 or is affected by some pulse during this time (I nuclei).

The preparation period often starts with a delay, which allows the spin system to relax back into its equilibrium state. The preparation period is usually terminated by a pulse or by a series of pulses that bring the spin system into a defined nonequilibrium state, which is the initial state for the subsequent evolution. The terminating pulse does not have to be a rf pulse. In Fig. 2.4b there is an example in which the nonequilibrium state is formed by a pulse of light that is used to initiate a chemical reaction.

The evolution period, like the detection period, cannot be omitted from the pulse sequence for the measurement of a 2D spectrum. Depending on the experimental arrangement, the spin system can be subjected to different conditions during the evolution period. The conditions chosen for the evolution period determine how the spin system evolves from the initial nonequilibrium state during t_1. Thus the physical conditions during the evolution period decide what sort of information will be encoded along the direction of the f_1 axis in the resulting 2D spectrum. During the evolution period the spin system can be subjected, for example, to rf pulses, or it can be irradiated continuously. Theoretically, these influences on the spin system are described by the Hamiltonian $\mathcal{H}^{(1)}$. The final state of the spin system at the end of evolution period depends not only on this Hamiltonian (i.e., on the physical conditions during t_1) and on the initial state, but also on the elapsed time t_1.

The mixing period separates the detection and evolution periods. The events contained in the mixing periods vary. In the Jeener experiment (Fig. 2.4a) the mixing period consists of a single mixing pulse, in other

[†]A spin system that consists of two nuclei with different chemical shifts is referred to as an "IS" spin system. In the heteronuclear case (e.g., ^1H and ^{13}C in $^1H^{13}CCl_3$), we shall consistently label as "I" the nucleus irradiated by the decoupler channel and as "S" the nucleus detected in the observation channel.

experiments the mixing period contains several pulses and delays (Fig. 2.5c), and in yet other experiments the mixing period is absent (e.g., in measurements of resolved spectra, Fig. 2.4e). The mixing period must be designed to ensure that the information about the evolution of the spin system during t_1 is passed on to the detection period in a suitable form. The evolution of the spin system during t_1 imprints the spin system in the detection period. For example, in heteronuclear measurements, this means that the mixing period must contain a combination of rf pulses that encode the information about the evolution of spins of one type, I spins (e.g. ^1H) into some property of the other type of spins in the system, S spins (e.g., ^{13}C). The information about the spin system evolution is found in the amplitude or phase of the spectra after the first Fourier transform. (For a discussion of amplitude or phase modulation, see Section A.7.) From this description it might appear that the period should be called an information transfer period. The term "mixing period," however, is closer to the actual physics of information transfer. The mixing pulses achieve information transfer by creating a new population distribution of spins in the energy levels and by forming new coherences (i.e., coherent motions of spins in various pairs of energy levels).

The detection period always starts with switching on (opening) the spectrometer receiver (in the observation channel) and ends by switching the receiver off after the detection time t_2 has elapsed. The FID that is obtained will depend, of course, on the conditions that are present during the detection period. Different FIDs and of course different spectra are obtained if the spin system is irradiated (decoupled) during t_2, or if the magnetic field is (or is not) homogeneous. (Figure 2.4d shows an example of using a nonhomogeneous magnetic field for obtaining a spatial image of an object, which is the basic experiment of NMR tomography.) The effects of these factors are described by the Hamiltonian of the system in the detection period $\mathcal{H}^{(2)}$.

In comparison to evolution time t_1, which is systematically varied during a 2D experiment, the detection period t_2 has an identical length t_2^{max} during the entire experiment. Only the maximum evolution time t_1^{max} has the same role in 2D spectra in the f_1-axis direction as the length of detection period t_2^{max} has in the f_2 direction.

In the same way that the length of the detection time (acquisition time) determines the achieved resolution in 1D spectroscopy, the t_1^{max} and t_2^{max} times determine the resolution achieved along the f_1 and f_2 axes, respectively. With longer maximum t_1 and t_2 times, narrower 2D peaks are obtained (Section 2.5).

The set of all FIDs, that is, the FIDs obtained for all evolution time t_1 values, represents the function $s(t_1, t_2)$. From the mathematical point of view, it is immaterial which Fourier transform is carried out first and which Fourier transform is second (i.e., whether the function is first transformed with respect to the t_1 or the t_2 variable). Usually, it is more economical to proceed as described for the Jeener experiment (Fig. 2.2), that is, to transform individual FIDs into interferogram spectra (a transform with respect to t_2) and then to transform the interferograms into 2D spectra (or, more aptly, into sections of the 2D spectrum).

Any 2D spectrum, as a function of two independent variables, $S(f_1, f_2)$, is a general surface in three-dimensional space. Hence we have a problem of how to present such a surface on a two-dimensional sheet of paper so that not only does the presentation show the characteristic and desired features, but also so that quantitative data can be obtained from the spectra. In the spectrometer (or in its computer) the signal dependences on time $s(t_1, t_2)$ and on frequency $S(f_1, f_2)$ are stored as large data matrices. The row and column of the matrix element correspond to the two variables or coordinates, either to t_1 and t_2, or to f_1 and f_2, and the value of the matrix element is the value of the signal. (The row and column indices are related to the address of the memory location that contains the signal value in digital form.) From these complete data, the computer can create various presentations or graphs according to our needs. Several modes of presentation of 2D spectra are used in the literature; they are reviewed in the next section.

2.2. THE APPEARANCE OF 2D NMR SPECTRA

Undoubtedly, the technical means for presentation of 2D NMR spectra will continue to improve. Newer and faster plotters, and displays with better resolution, will continue to appear. It is not very likely, however, that the means for presentation of 2D spectra will change fundamentally in the near future, although black-and-white pictures might give way to color.

The 2D spectra or parts thereof are customarily presented graphically in one of two forms: stacked-trace plots and contour plots. A stacked-trace plot is obtained by drawing a series of spectra $S(f_2)$ one behind the other in order of increasing frequency f_1 [or by plotting $S(f_1)$ spectra in order of frequency f_2]. The spectra are arranged in the figure in such a way that a three-dimensional impression is created (Fig. 2.6). This impression is enhanced by "whitewashing," that is, by leaving out those parts of the traces that have a lower intensity than previous traces and therefore would not be seen in space (Fig. 2.7).

The three-dimensional feeling is further enhanced by a skewed presentation, with each trace starting not only a few millimeters above the preceding one but also shifted slightly to the left or right (see the example in Fig. 1.3). One common danger when using whitewashed presentation methods is that peaks which are "hidden" behind peaks in preceding traces might escape our

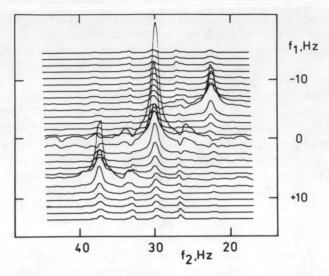

Figure 2.6 Graphical presentation of a 2D spectrum in the form of a stacked-trace plot.

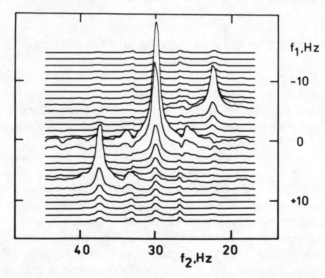

Figure 2.7 Graphical presentation of a 2D spectrum in the form of a whitewashed stacked-trace plot; same spectrum as in Fig. 2.6.

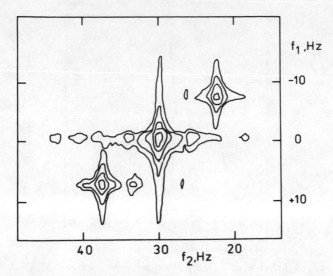

Figure 2.8 Contour plot of a 2D spectrum; same spectrum as in Figs. 2.6 and 2.7.

attention, as would also any peak with a negative intensity. The whitewashed skewed stacked-trace plots can give an attractive appearance, but finding the exact location of a peak (i.e., its f_1 and f_2 coordinates) is difficult, as is a precise measurement of the distance between two peaks in such graphs. Plotting is also time consuming, and very often the attractive pictures that one sees published are the result of several trials (and tribulations).

Contour plots depict the intensity distribution in the 2D spectrum just as contour lines indicate altitudes in geographical maps. The contour lines connect points (f_1, f_2) that have the same signal intensity. Figure 2.8 shows such a plot for the spectra that were presented in Figs. 2.6 and 2.7. Plotting is usually faster than in the case of stacked-trace plots, but problems arise when both low- and high-intensity peaks are present in the spectrum. If the minimum contour level is chosen to show the presence of low-intensity peaks, the high-intensity peaks extend over a large area, and contours of spurious signals or noise crowd the spectrum. The precision in determining the position of a peak depends on the diameter of the highest contour level surrounding the peak.

Cross sections are most suitable for obtaining quantitative data from 2D spectra. Which sections bear relevant information depends on the type of 2D spectrum, but it is often necessary to have sections through the spectrum parallel with the f_1 or f_2 axis. In the case of homonuclear J-resolved spectra, however, sections inclined at 45° to the f_1 axis carry important information.

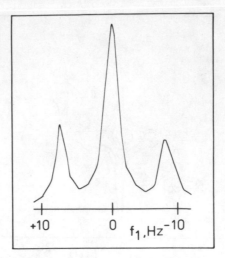

Figure 2.9 Integral projection of a 2D spectrum; same spectrum as in Figs. 2.6 through 2.8; projection along the f_2 axis onto the f_1 axis.

Plotting only the relevant sections saves a lot of time in comparison with plotting the entire stacked-trace spectrum.

Projections of spectra along certain directions serve a similar purpose to cross sections, and in certain other types of spectra the projections have distinct advantages that will be discussed later. A projection can be obtained as a "skyline" projection [27] or as an integral projection. The skyline projections show the maximum (or minimum) intensities in the 2D spectrum in the direction of the projection. Integral projections show an intensity integral (summation) through the spectrum in the direction of the projection. An example of an integral projection is shown in Fig. 2.9.

Before discussing the shape of 2D peaks, let us recall some phase properties of conventional 1D NMR spectra. A Fourier transformation of a FID usually yields a spectrum in the form shown in Fig. 2.10a. (The distorted lineshapes are apparent only if the spectrum is presented in the phase-sensitive mode, which is the more-or-less standard mode for presentation of 1D spectra.) The distorted lineshape is a consequence of nonideal behavior of rf circuitry and of receiver "dead time," which distorts the FID. Hence FT does not produce a pure absorption lineshape but instead gives lines with a shape that is a mixture of absorption and dispersion lineshapes. This distortion is easily removed by a mathematical procedure known as phase correction (see Section A.2). After phase correction, all lines have the shape of absorption lines (i.e., the lines are symmetrical with respect to the

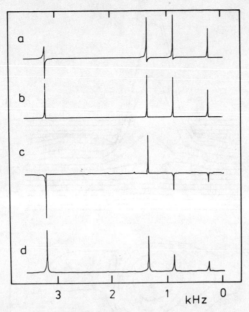

Figure 2.10 One-dimensional ^{13}C NMR spectra of 2-butanol: (*a*) spectrum after FT without phase correction; (*b*) the same after phase correction; (*c*) APT spectrum; (*d*) APT spectrum in absolute mode presentation. The unequal intensity of the lines is due to fast pulse repetition.

axis passing through the maximum and perpendicular to the frequency axis). Also, all peaks have the "correct" (i.e., positive) intensity (see Fig. 2.10*b*).

Some types of 1D measurements, however, yield spectra with absorption lineshapes but with some lines being inverted (i.e., they have negative intensities). Such lines are encountered in measurements of relaxation times, CIDNP or APT spectra (Fig. 2.10*c*), and others. As we usually associate "normal" or "positive" intensity with absorption of radiation, the inverted lines suggest an emission of rf radiation. Indeed, the inverted lines do come from inverted energy-level populations in measurements of CIDNP spectra or relaxation times.

When inverted lines are present in the spectrum, phase correction cannot bring all the lines into an absorption lineshape with positive intensities. All lines have positive intensities and no phase correction is needed if we replace the common phase-sensitive mode of presentation with either an absolute-value mode or a power spectrum mode. The former replaces the signal intensity with its absolute value in the spectrum, and the latter uses the second power of the signal intensity. As is apparent from Fig. 2. 10*d*, the line position is not altered by the change in the mode of presentation, but some

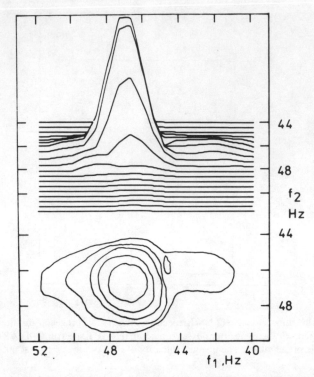

Figure 2.11 Example of a real 2D peak with a shape close to Gaussian: upper part, stacked-trace plot; lower part, contour plot.

information that was present in the spectrum is lost in this mode (e.g., in phase-sensitive ^{13}C NMR APT spectra, the lines with negative intensities correspond either to methyl or methine carbons, and lines with a positive intensities are due to methylene carbons). Also, since the intensity ratios in power spectra are raised to the second power, the differences between strong and weak lines in the spectrum are emphasized. These modes also have broadened lines at the base of the spectrum.

Surprisingly, the phase properties of 2D spectra are much more complex. One might expect the 2D peaks to have a shape formed by rotation of a 1D absorption line about its axis of symmetry. This is true, however, only for some 2D peaks (Gaussian peaks) that have circular (or elliptical) contour lines (Fig. 2.11). More common are Lorentzian peaks with their more complicated shape, as shown in Fig. 2.12. The 2D Lorentzian peak does not possess rotational symmetry, and the contours follow the characteristic shape of a four-pointed star.

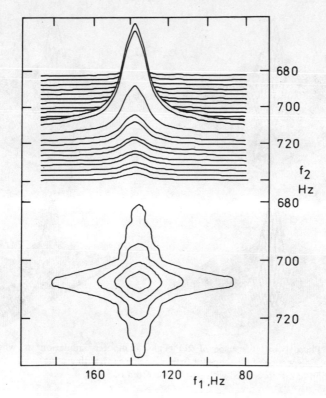

Figure 2.12 Example of a real 2D peak with a Lorentzian shape: upper part, stacked-trace plot; lower part, contour plot.

Only a few 2D NMR experiments yield pure absorption peaks with a Gaussian or Lorentzian lineshape in the phase-sensitive mode. Other experiments can be modified to yield such spectra, but only at the expense of measuring time and more complicated computations (see Appendix Section A.7).

Very frequently, however, it suffices to perform 2D experiments that produce lines with a "phase twist." The complicated phase twist lineshape is shown in Fig. 2.13. A vertical section through a peak that is far from resonance shows a dispersion mode shape; a section at resonance shows a pure absorption line shape. Sections in the intermediate regions show a mixture of the two shapes.

The contribution from the dispersion lineshape has some undesirable effects on the spectrum. First, the dispersion mode is broad, and the wings have a nonzero intensity even if far from the resonance. For this reason, the

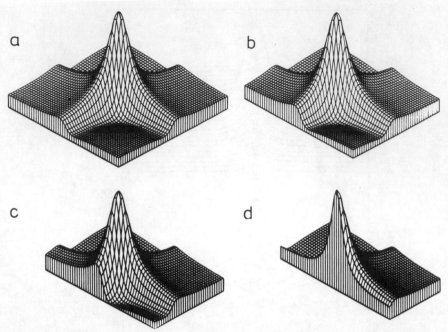

Figure 2.13 "Phase twist" lineshape of 2D peaks: computer simulations in which vertical sections show a dispersion mode shape far from resonance (*a*), with gradual progression toward pure absorption (*d*). Reproduced with permission from ref. 13.

peaks are much broader than would correspond to an absorption lineshape. The broadening is most apparent at the baseline. Wings from strong lines can overlap with weak lines, and the weak lines can therefore escape our attention. The wings have intensities of opposite signs on the two sides of the resonance, and this can cause problems when the lines overlap; wings from different lines can cancel each other. When spectra are projected on one of the f_1 or f_2 axes, the dispersive wings cancel each other in an integral projection, and the projection shows a pure absorption peak. On the other hand, a projection along the line that encompasses a 45° angle with the f_1 axis has zero intensity over the entire spectrum unless an absolute value or a power spectrum is used for the projection (or the spectrum must be measured by a method yielding a pure absorption lineshape). Otherwise, the part with negative intensity will exactly cancel the part with positive intensity, as explained above in connection with overlapping wings.

The discussion above explains why it is not possible to give a simple straightforward recipe of how to produce graphical presentations of 2D spectra. The correct choice depends on many aspects: some of them are of a fundamental nature (since the type of 2D NMR spectroscopy or the pulse sequence used will determine the lineshape); other choices will be dictated by the particular presentation needs and various circumstances peculiar to the experiment. It is also necessary to consider these needs in view of the possibilities of the hardware and software of the given spectrometer.

More often than not, it is necessary to look for the best presentation by a trial-and-error approach. Frequently, just turning the spectrum by 90° reveals signals that were originally hidden. The illustrations in publications are usually the results of many hours spent in a search for the best presentation. It is likely, however, that in the future fewer and fewer complete 2D spectra will be accepted for graphical presentation in journals. Most of the spectral features can be adequately demonstrated in sections, which not only require less space but are easier to reproduce in good quality.

2.3. CLASSIFICATION OF EXPERIMENTAL METHODS AND LINES IN THE SPECTRA

The scores of pulse sequences described in the literature [17] serve a variety of needs of 2D NMR spectroscopy. They include some very useful pulse sequences as well as modifications that are supposed to improve some feature of the measurement or of the resultant spectrum. Some pulse sequences obviously have only a limited use, since they have been designed to serve a special purpose (e.g., to solve problems with a particular set of coupling constant values), but only time will tell which of the many pulse sequences will remain as a standard part of the 2D arsenal and which will be forgotten. This textbook cannot give an exhaustive description (or even merely a list) of all the 2D NMR pulse sequences, and we shall limit ourselves to providing only a description of the functions of a few of the most fundamental (usually the simplest) pulse sequences and their modifications. Even this limited goal, however, cannot be accomplished without some classification system for 2D pulse sequences.

A division of 2D NMR methods into homonuclear and heteronuclear methods (see Figs. 2.4 and 2.5) is not very significant. It does, of course, divide the methods into two groups: the methods in the heteronuclear group can be described and explained with sufficient precision by the semiclassical vector model; the methods utilized for the homonuclear experiments, however, are not all adequately described by this model. Moreover, this division is no longer useful from an experimental point of view since all commercial spectrometers on which 2D NMR spectra can be measured are equipped for the measurement of both homonuclear and heteronuclear 2D spectra.

A more fundamental division of the methods examines the presence or absence of a mixing period in the pulse sequence. In the mixing period, information is transferred from one system (which has evolved during t_1) to another system (which will be measured during the detection period), and the peaks in the resulting 2D NMR spectrum correlate f_2 frequencies with f_1 frequencies. For this reason the methods that have a pulse sequence containing a mixing period are referred to as "correlation methods" and the spectra as "correlated" 2D NMR spectra. If the pulse sequence does not contain a mixing period with pulses, only the same system can be followed in both the evolution and detection periods. This is useful only if that system is exposed to two different physical conditions in the two periods; otherwise, the results of the measurement would be trivial. No information transfer is needed since the system "remembers" its own past; that is, it "remembers" during time t_2 what has happened to it earlier during time t_1.

Different physical conditions during the evolution and detection periods can be chosen so that the conventional 1D NMR spectrum measured during t_2 is spread out into a second dimension according to some selected NMR parameter (e.g., according to the values of the heteronuclear spin-spin coupling constants pertaining to the lines observed in the conventional spectrum). This spreading of a "linear" 1D spectrum into a two-dimensional surface increases the spectral resolution as discussed in Chapter 1, and hence the spectra (and the methods) are labeled "resolved". Sometimes the label is specified further by stating the NMR parameter according to which the spectra were spread out (e.g., J-resolved). Generally, the axes of 2D resolved spectra may indicate various combinations of NMR parameters (chemical shifts, coupling constants) depending on the physical conditions applied to the spin system during times t_1 and t_2.

It should be noted that correlated spectra also increase the spectral resolution of NMR spectroscopy, and the two accidentally identical chemical shifts discussed in Chapter 1 could be differentiated by either correlated or resolved spectra.

The categorization above is rather abstract. Let us briefly consider two concrete examples that will be explained later in much more detail.

1. J-resolved homonuclear proton 2D NMR spectra are measured by the pulse sequence shown in Fig. 2.4e. The 180° pulse in the middle of the evolution period eliminates the effect of chemical shifts on the evolution of the spin system. Thus, at the end of the evolution period the nuclear spins are oriented as if they had evolved only under the influence of spin-spin coupling (and all chemical shifts were zero) for the entire duration of t_1. Therefore, the lines in the spectra obtained by the first Fourier transform, $S(t_1, f_2)$, are affected only by the structure of the spin-spin multiplets. This influence is

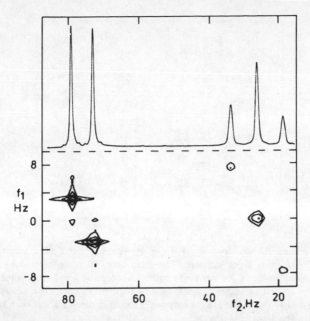

Figure 2.14 Homonuclear J-resolved 2D NMR spectrum (^{1}H) of 2-butanol: lower part contour plot of the 2D spectrum; upper part, conventional 1D ^{1}H NMR spectrum. Only a part of the spectrum is shown, the doublet and triplet are due to the protons of the two methyl groups.

more clearly seen from the 2D spectra after the second FT. A contour plot of such a spectrum is shown in Fig. 2.14. The f_2 coordinate of a peak in this spectrum corresponds to the position of a line in the conventional 1D spectrum, and the f_1 coordinate corresponds to the position of the line within its multiplet, that is, to the position of the line with respect to the chemical shift of the multiplet.

2. Correlated homonuclear 2D NMR spectra are measured by the two-pulse sequence described by Jeener (Figs. 2.2a and 2.4a). The mixing period of this pulse sequence consists of a single mixing pulse only. The nuclear spins that were on energy levels m and n during time t_1 moved coherently so that they gave rise to a magnetization component that precessed with frequency f_{mn} during the t_1 period (although we have not detected a FID during this period). The mixing pulse moves these nuclei to other levels, and they contribute to magnetization components that are due to the nuclei being on these new levels. What we detect during t_2 depends on how many nuclear spins are transferred from levels m and n to levels k and l, and how much these transferred nuclei contribute to the magnetization component precessing with frequency f_{kl}. The results thus depend on two factors: on evolution

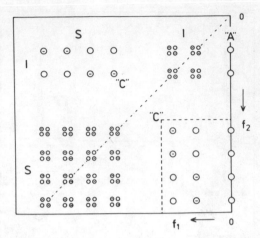

Figure 2.15 Correlated 2D spectrum of a I_3S spin system; schematic phase-sensitive contour plot with negative intensity signs indicated. The main diagonal is represented by the broken (—·—·) line; "A," axial peaks; "C," cross-peaks with an absorption lineshape, diagonal peaks of the S and I nuclei around the main diagonal have a dispersion shape; cross-peaks in the box separated by the dashed (– – –) line are the only peaks found in heteronuclear correlated spectra when the signal of the S nuclei is detected.

time t_1 and on the relationships among the energy levels m, n, k and l. The latter factor is related to the interpretation of the spectra (see Section 4.1.2). It should suffice for now to note that the nuclei are transferred only between energy levels within one coupled spin system; no transfer is possible between energy levels that belong to different spin systems that are not spin-spin coupled. Within the coupled spin system, the gain of magnetization with precession frequency f_{kl} (or the intensity gain of the line with this frequency after the first FT) varies periodically with time t_1 and also with frequency f_{mn}, that is, with the frequency of the line from which the intensity gain has originated (and which was not detected during t_1). This is the frequency with which the intensity of line f_{kl} varies in the spectra $S(t_1, f_2)$ measured for different t_1 times. This frequency is apparent in the interferogram corresponding to $f_2 = f_{kl}$. After the second FT, the f_1 coordinate of the 2D peak with $f_2 = f_{kl}$ is equal to f_{mn} in the correlated 2D spectrum.

From the description of a correlated 2D NMR experiment above, it should be clear that three different types of peaks can be found in homonuclear correlated spectra:

1. Peaks that correspond to a line with the same frequency f_{mn} in both time periods. These diagonal peaks with coordinates $f_1 = f_2$ are due to

nuclei that have remained on the same energy levels during both the evolution and detection periods.

2. Peaks that correlate lines with different frequencies f_{mn} and f_{kl} are called cross-peaks.

3. Peaks to which no information from the t_1 period has been transferred. These peaks lie on the axis $f_1 = 0$ and are therefore called axial peaks. Since axial peaks bring no new information, efforts are usually expended to eliminate the presence of axial peaks by experimental techniques such as phase cycling.

The relative positions of the three types of peaks are shown schematically in Fig. 2.15. From the definitions above it is clear that diagonal peaks cannot be observed in heteronuclear correlated spectra, but the positions of the cross-peaks in the directions of both axes provide sufficient information to determine the position (chemical shift) of diagonal peaks of both types of correlated nuclei. Obviously, this peak classification does not apply to resolved spectra that are simple enough not to require such a classification.

2.4. PHYSICAL DESCRIPTION OF THE SIMPLEST 2D NMR EXPERIMENTS

A rigorous analysis of pulsed NMR experiments requires the use of density matrix formalism, which is still rather foreign or "user unfriendly" to the chemist who needs 2D NMR to analyze a sample and wants to have some idea about the basis of the measurements. (A full understanding of the physics and the physical mechanisms that operate during a 2D NMR experiment is not an essential prerequisite for successful use of the technique, but it is very helpful both in setting up the experiments and in analyzing the results.)

Some very simple heteronuclear 2D NMR experiments can be described quite adequately by a semiclassical vector model. Despite its limitations, the vector model is a valuable tool to gain physical insight into the behavior of spin systems during NMR experiments. It provides a clear picture of the origin of 2D spectra, and it can be used in a search for improvement of the pulse sequence or to delineate the possibilities and limitations of a particular method. The physical picture formed with the help of the vector model also played an important heuristic role in the design of 2D NMR pulse sequences [14].

We shall use the vector model here to elucidate the mechanism of the simplest variants of experiments to measure correlated and resolved heteronuclear 2D NMR spectra. Those readers who are not familiar with the description of spin system behavior by the vector model are advised to read Appendix Sections A.1 to A.4 before continuing with this section. The Appendix explains the fundamental features of the vector model and how it can be used to describe the motion of spin systems.

We shall begin our consideration with the simplest heteronuclear spin system IS. After the behavior of this system is understood, we will discuss more complicated cases such as I_2S and I_3S spin systems. Since the measurement of resolved spectra does not involve a mixing period, it is conceptually simpler and easier to understand the measurement of resolved spectra than correlated spectra. For this reason we begin with a variant of the gated decoupler method for measurement of heteronuclear resolved 2D spectra. (Other methods are considered in Section 3.1.) The gated decoupler method, which is often referred to as the Müller–Kumar–Ernst (MKE) experiment because of the names of its inventors [28], is particularly simple; it is one of the oldest 2D NMR techniques.

2.4.1. Heteronuclear Resolved Spectra

The method proposed by Müller et al. [28] in 1975 uses the pulse sequence shown at the top of Fig. 2.16. Let us see how it acts on an IS spin system (e.g., chloroform, $^1H^{13}CCl_3$).

The 90° pulse turns the macroscopic magnetization M^S of the S nuclei (i.e., ^{13}C in our chloroform example) from its equilibrium orientation into the x', y' plane of the rotating coordinate system, which rotates with the frequency f_r of the observing transmitter. Immediately after the pulse (the instant when $t_1 = 0$), the magnetization begins its rotation around the direction of the magnetic field $\vec{B}_0$. Because of the two possible states of the I nuclei,* the magnetization M^S is composed of two components of the same magnitude: M^{S+} and M^{S-}. The M^{S+} component is a vector sum of the magnetic moments of those spins S which are in the molecules that have their spin I (^1H in our example) in the $< + >$ state. Analogously, the M^{S-} component comes from molecules with their spin I in the $< - >$ state. Both components rotate during t_1. (More correctly, they precess, but at the moment we ignore relaxation in our considerations.) Both components rotate with the frequency f_S, which corresponds to the chemical shift of nucleus S [i.e., they rotate with the frequency $f_S - f_r$ in the rotating frame (of reference)]. Their vector sum, M^S, the total magnetization of nuclei S, also rotates with this frequency. In addition to this rotation, however, the two components also rotate with respect to each other with a frequency equal to

---

*The two states correspond to the magnetic quantum numbers $m = +\frac{1}{2}$ and $m = -\frac{1}{2}$, which are abbreviated here as $< + >$ and $< - >$.

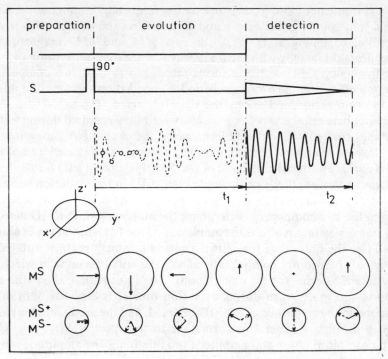

Figure 2.16 Müller–Kumar–Ernst [28] experiment; IS spin system with the I nucleus decoupled and the S nucleus detected. Upper, pulse sequence and NMR signal; the FID detected is shown by a solid line, the FID that is not directly detected during the evolution period is shown by a dashed (–––) line. Lower, M^S magnetization and its M^{S+} and M^{S-} components in x', y' plane at selected times of the evolution period marked by circles on the undetected FID.

the spin-spin coupling constant J_{IS} (hertz). This additional rotation causes periodic changes in the magnitude of M^S, as indicated at the bottom of Fig. 2.16. With the two rotations combined, the M^{S+} component rotates with the frequency $(f_S - f_r) + J_{IS}/2$, and the M^{S-} component rotates with the frequency $(f_S - f_r) - J_{IS}/2$. If we detected the FID during the evolution time, it would have the complicated shape indicated by the dashed line in Fig. 2.16. (This type of complicated FID is detected when the 1D spectra of S are measured without decoupling of the I nuclei, for example, in measurements of proton-coupled ^{13}C NMR spectra.) A Fourier transformation would produce a doublet centered at $f_S - f_r$ with a line separation equal to J_{IS} (the doublet of the chloroform carbon in our example). In the 2D NMR measurements discussed, no FID is detected during the evolution time; instead, the evolution of the spin system is interrupted after time t_1 by switching on the decoupler. Decoupling (i.e., the irradiation of the I nuclei with a rf frequency close to their resonating frequency) eliminates the effect of spin-spin coupling (between I and S nuclei) on the motion of S nuclei. Therefore, beginning at $t_2 = 0$, the two M^{S+} and M^{S-} magnetization components rotate only with the frequency $f_S - f_r$; they both rotate with the same frequency. Of course, the angles through which the two components have rotated during the evolution period is encoded on their motion during the detection period and affects the signal detected. The angle (the phase difference) between the two components, which they acquired during time t_1 when they rotated with different frequencies, is not changed during time t_2, and thus it determines the amplitude of the FID detected (amplitude modulation). The initial amplitude of the directly detected FID (at $t_2 = 0$) is the same as the amplitude of the undetected FID in the evolution period at the instant when the evolution has been interrupted; thus the phase difference between the two components determines the amplitude of the FID detected in the same way that it affects the amplitude of the FID during the evolution period (see the bottom of Fig. 2.16). The angle, which has been rotated by the sum of the two M^S components, affects the initial phase with which the detected FID begins (phase modulation). The information about the evolution of the spin system during evolution time t_1 is encoded both in the phase and in the amplitude of the FID detected, and the modulation is passed through the first Fourier transformation to the spectra, $S(t_1, f_2)$, which exhibit combined phase and amplitude modulation. (The amplitude modulation carries the information on the motion of the components or the splitting of the multiplet, and the phase modulation carries information about the chemical shift of the multiplet.) The second FT of the interferograms [remember that interferograms are obtained from the spectra, $S(t_1, f_2)$, by data transposition so that a second FT with respect to time t_1 can be performed] produces a 2D spectrum (see Fig. 2.17) that has a doublet in the direction of the f_1 axis, as found in spectra measured without decoupling. The f_2 coordinates of the two peaks are the same; they correspond to the chemical shift of the S nucleus. In the direction of the f_2 axis, the spectrum has the appearance of a 1D spectrum measured with decoupling. (Information about evolution was encoded only into those interferograms with a f_2 frequency close to the frequency of signal in the decoupled spectrum.)

The resolved 2D spectrum obtained in the manner described above can be viewed from two different perspectives. On the one hand, looking from the f_1 axis, that is, from the point of view of the spectrum measured without decoupling, spreading the spectrum into the f_2 axis has the effect of resolving the spectrum according to the chemical shifts ("chemical shift resolved spectra" or "δ-resolved"). On the other hand, looking at the 2D spectrum from the point of view of the decoupled spectrum (i.e., the view from the f_2 axis), spreading the spectrum into the f_1 dimension introduces the effect of

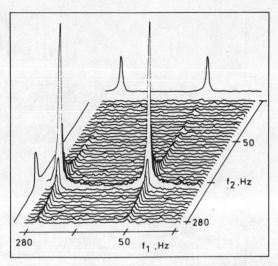

Figure 2.17 Resolved 2D NMR (^{13}C) spectrum of chloroform measured by the MKE experiment; eight transients for each t_1 value, 128 increments of t_1, acquisition time $t_2 = 0.27$ s, spectral width 470 Hz in both the f_1 and f_2 axes, the FID data matrix 256 × 128, zero-filled to 256 × 256 before transforms, total measuring time 1 h 18 min; absolute-value presentations, no digital filtering; $(f_S - f_r) = 170$ Hz and $J_{HC} = 208$ Hz; the spectra, which were plotted along the f_2 axis with proton decoupling and along the f_1 axis without proton decoupling, were obtained in separate 1D measurements; that is, they are not projections on the respective axes.

heteronuclear spin-spin coupling, and the spectrum can be considered resolved according to spin-spin coupling (J-resolved). Both terms are found in the literature.

The mechanism described can easily be extended to more complicated spin systems such as I_2S and I_3S. In these systems, the total magnetization, M^S, of the S nuclei is composed of a greater number of components. In the I_2S system we must consider three components, M^{S+}, M^{S0}, and M^{S-}, which rotate during the evolution period with the frequencies of triplet components, that is, with the frequencies $[(f_S - f_r) + J_{IS}]$, $[(f_S - f_r)]$, and $[(f_S - f_r) - J_{IS}]$, respectively (in the rotating frame of reference). Similarly, the M^S magnetization in the I_3S system consists of four components, $M^{S+3/2}$, $M^{S+1/2}$, $M^{S-1/2}$, and $M^{S-3/2}$, which rotate with frequencies $(f_S - f_r) + J_{IS} \Sigma m_I$, where Σm_I is the sum of the magnetic quantum numbers of the I spins in the particular state of the whole system. The sums are indicated as the index at the corresponding magnetization components. All these components rotate during the evolution period, and their vector sum would produce a complicated FID if it were detected. After the decoupler is switched on, the motion of the components is simplified. All components continue to rotate

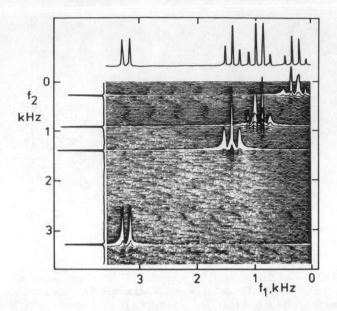

Figure 2.18 Stacked-trace plot of a resolved 2D NMR (^{13}C spectrum of 2-butanol in hexadeuteriobenzene measured by the MKE pulse sequence; four transitions acquired for each t_1 value, 128 increments of t_1, acquisition time $t_2 = 0.07$ s, spectral width in both axes 3600 Hz, measuring time 35 min, FID matrix 256 × 256, no digital filtering, absolute-value presentation, stacked-trace plot with whitewash. The 1D spectra added at the top of the figure were obtained by independent measurements.

with the same frequency, $(f_S - f_r)$, and a FID corresponding to a decoupled spectrum is detected. Its phase and amplitude, however, depend again on time t_1 and the frequencies that the components have had during this time. An example of a 2D resolved spectrum of 2-butanol (with two I_3S, one IS, and one I_2S spin systems) is shown in Fig. 2.18 in stacked-trace form and in Fig. 2.19 in a contour plot presentation, respectively.

Obviously, the method described spreads the multiplets of a linear spectrum into the two-dimensional plane and thus enhances spectral resolution. (Without such spreading the 1D spectra of large molecules measured without decoupling are difficult to analyze because of severe multiplet overlap.) However, the MKE method suffers from several drawbacks (e.g., low sensitivity, and the resolution in the f_1 axis is limited by magnetic field inhomogeneity and large requirements on computer memory size). Therefore, several improvements have been suggested. The improved methods, as well as other methods for measurement of resolved spectra, are discussed in detail in Section 3.1.

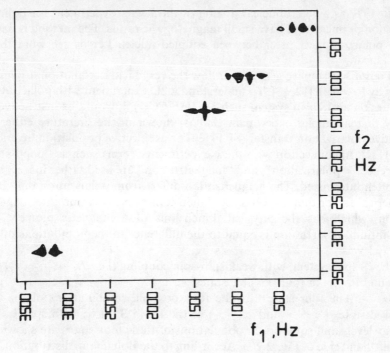

Figure 2.19 Contour plot of a resolved 2D NMR (^{13}C) spectrum of 2-butanol; same spectrum presented in stacked-trace plot form in Fig. 2.18.

A resolved 2D NMR spectrum shows the relationship (correlation) between the lines in the decoupled spectrum and in the coupled spectrum (a monoresonance spectrum). For this reason, we can also regard resolved spectra as a special case of correlated 2D NMR spectra, which are measured with exclusion of the mixing period.

2.4.2. Heteronuclear Correlated Spectra

The 2D NMR spectra that are conceptually the simplest and most suitable for an introduction to the field of correlated 2D NMR spectroscopy correlate the signals of two 1D NMR spectra of two different nuclei and do so on the basis of spin-spin coupling between these nuclei. Very frequently (although not very accurately), such 2D spectra are called "chemical shift correlated spectra." Since chemical shift correlated spectra have a broad "spectrum" of applications and great experimental variability, they are perhaps the most useful of all the various types of 2D NMR spectra. Although there are many methods, the basic one had already been proposed by Maudsley and Ernst

[29] in 1977 as a heteronuclear analog of the Jeener experiment for indirect detection of nuclei that have small magnetogyric ratios. The method is based on a polarization transfer between coupled nuclei. Let us see what these terms mean.

In our discussion we generally follow the very lucid explanation originally given by Freeman [12, 13]. To understand a 2D experiment with polarization transfer on an IS spin system such as (^{1}H^{13}CCl$_3$), it would be instructive to study first a simpler experiment that is known in the literature either as selective polarization transfer (SPT) [30] or as selective population inversion (SPI) [31]. In connection with these experiments, terms such as "population difference," "polarization", and "magnetization" are used rather loosely and are often intermixed. The magnetization differs from polarization only by a numerical factor that converts the dimensionless polarization into magnetization, which has the physical dimensions of a magnetic moment. At equilibrium, polarization is equal to the difference in the populations of the energy levels.

The IS spin system, with weak spin-spin coupling (i.e., $J_{IS} \ll f_I - f_S$) can be found in one of four possible states: $< + + >$, $< + - >$, $< - + >$, and $< - - >$. The four states are the four possible combinations of the two allowed states ($< + >$ and $< - >$) of the I and S nuclei with spin $\frac{1}{2}$. The energy levels and equilibrium populations of these four states are shown on the left-hand side of Fig. 2.20a. According to the Boltzmann distribution, the difference in level populations is proportional (to a first approximation) to the energy difference between the two levels. Since the difference in energy between two levels is proportional to the frequency of the transition connecting the two levels (Planck's law), the population difference (and hence the signal intensity) is also proportional to the frequency of the transition, i.e., to the frequency of the corresponding NMR line.

In our example, ^{1}H^{13}CCl$_3$, the differences in populations of the levels connected by proton transition from the $< + >$ state to the $< - >$ state are approximately four times larger than the differences in populations of the levels connected by carbon transition (remember that the magnetogyric ratio of carbon is about one-fourth that of the proton; when a proton resonates at 100 MHz, carbon resonates at 25 MHz). Proton transitions connect the $< + + >$ → $< - + >$ and $< + - >$ → $< - - >$ states, while carbon transitions connect the $< + + >$ → $< + - >$ and $< - + >$ → $< - - >$ states. The differences in the populations of the levels are reflected in the intensities of the lines in spectra of the I and S nuclei (the right-hand side of Fig. 2.20). We shall label the lines in the spectrum in agreement with the preceding section: the M^{S+} line (magnetization) in the spectrum of the S nuclei is due to S nuclei from the IS molecules (spin systems) that have their spin I in the $< + >$ state. The spectral line is caused by the $< + + >$ →

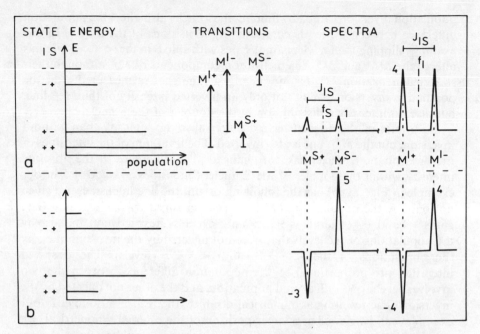

Figure 2.20 An IS spin system with $J_{IS} > 0$: (*a*) at equilibrium; (*b*) after selective inversion of the M^{I+} line. Schematically, the spectra correspond to the $^1H^{13}CCl_3$ molecule; the numbers give the relative line intensities considering population differences only.

$< + - >$ transition. The line appears in the spectrum as a Fourier transform of that part of the FID that is due to rotation of the M^{S+} magnetization. Other lines are analogously connected with the corresponding magnetization components.

Let us now turn to the SPI (or SPT) experiment. In the SPI experiment we subject the spin system at equilibrium to a selective 180° pulse. The selective pulse affects only one NMR line of the I nuclei (e.g., the M^{I+} line). [The selective pulse has the frequency of the M^{I+} line and such a low magnetic field induction that the 180° pulse has a duration of about 1 s (see Appendix Section A.4.1).] This pulse turns that M^{I+} magnetization component from the direction along the $+z$ axis to the opposite orientation along the $-z$ axis. The change in orientation corresponds to the inversion of populations of the $< + + >$ and $< - + >$ levels that are connected by the transition with the frequency of the M^{I+} line. (By population inversion, of course, we mean that the two levels exchange populations.) The populations of the levels after inversion of the M^{I+} line are depicted in Fig. 2.20*b*. Since the energy levels of the system are common for both the I and the S nuclei of the spin system, the population inversion caused by flipping the I nuclei also alters the population difference between the levels connected by transitions of the S spin. In our example, flipping the I nuclei in molecules with spin S in the $< + >$ state has altered the M^{S+} and M^{S-} magnetization components of M^S, although their sum, magnetization M^S, has not changed. The spectra measured after the population inversion exhibit not only an inverted intensity of the M^{1+} line, but also an intensity redistribution in the doublet of the S nuclei.

Both lines of the doublet have a larger absolute intensity than without inversion, but the M^{S+} line is also inverted. The total sum of the intensities of the lines of the S nuclei has not changed. The difference in the absolute intensity within the doublet of the S nuclei (intensities -3 and $+5$ in our example in Fig. 2.20b) has the following origin: the line intensities are given not only by the populations that have been changed by the inversion (i.e., gains $+4$ and -4 on levels $< - + >$ and $< + + >$, respectively), but also by the populations of the levels that were not affected by the inversion (i.e., by the populations of the $< - - >$ and $< + - >$ levels). The observed intensities are proportional to the population difference between the two involved levels, one with altered population, and the other not affected by the inversion. The resulting asymmetrical doublet (the -3, $+5$ doublet in our example) can be viewed as a super-position of the original symmetrical $+1$, $+1$ doublet without inversion and the "inverted" doublet (the -4, $+4$ doublet).

You can easily prove that a similar SPI experiment with the inversion of the other line of the I nuclei (M^{1-}) would produce a spectrum with a reversed intensity ratio in the doublet of the S nuclei. Obviously, inversion of one of the M^{1+} and M^{1-} lines will have the opposite effect on the spectrum of the S nuclei than will inversion of the other line. If we completely inverted both lines of the I nuclei (by using a nonselective pulse), both lines of S would be inverted, but no change would be detected in their relative intensities.

When the inversion of a line of the I nuclei is imperfect (e.g., because a pulse shorter than 180° is used), the population transfer is incomplete (the change in populations is smaller). The population transfer achieved is proportional to the z component of the partially inverted magnetization. In our example, if the selective pulse flip angle is α, the M^{1+} magnetization has its z component M_z^{1+} equal to $M^{1+} \cos\alpha$ after the pulse, and the intensity changes in the spectrum of the S nuclei show the same dependence on α.

Let us now refer again to the experiment of Maudsley and Ernst [29]. This experiment does not employ a selective pulse for population inversion, but it does use a nonselective 180° pulse divided into two 90° pulses. The interval t_1 between the two pulses acts as a frequency filter. Let us follow the motion of the M^{1+} magnetization during the experiment, as shown in Fig. 2.21.

Before the first nonselective 90° I pulse is applied, the orientation and

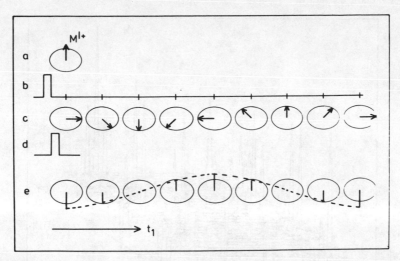

Figure 2.21 M^{1+} magnetization (IS spin system) behavior during the Maudsley–Ernst experiment [29]: (a) M^{1+} equilibrium magnetization before the first pulse; (b) the first nonselective 90° pulse; (c) M^{1+} magnetization in equidistant times t_1 after the first pulse; (d) the second nonselective 90° pulse; (e) the z-axis component, M_Z^{1+} of magnetization M^{1+} immediately after the second pulse, which also occurred in equidistant t_1 times.

magnitude of the M^{1+} magnetization component are given by the Boltzmann distribution. The magnetization points in the direction of the $+z$ axis (see the upper portion of Fig. 2.21). Immediately after the pulse (a $+x$ pulse that has its rf magnetic field in the direction of the $+x'$ axis, see Section A.3), the M^{1+} magnetization component is rotated into the x', y' plane and points in the direction of the $+y'$ axis. During the subsequent time t_1, the M^{1+} magnetization rotates with the frequency $(f_1 - f_r) + J_{IS}/2$ (hertz), just as it rotated during the evolution period in the MKE experiment (Fig. 2.16) that was discussed previously. The positions taken by the M^{1+} magnetization at selected moments t_1 are shown in Fig. 2.21c. (The M^{1+} magnetization rotates by 45° between the selected equidistant moments.) The second pulse, which is again an x pulse, rotates the y' axis component of the M^{1+} magnetization into the direction of the $-z'$ axis. The x' axis magnetization components are not rotated by the x pulse, and the z axis component is zero after the first pulse. In this way the second pulse completes the inversion of the M^{1+} magnetization that was begun by the first pulse. The extent to which the magnetization is inverted depends on the angle traveled by the magnetization during time t_1, since only the y' axis component of the magnetization is inverted. The $-z$ axis components after the second pulse are shown at the bottom of Fig. 2.21e for selected durations of t_1. The y' component varies

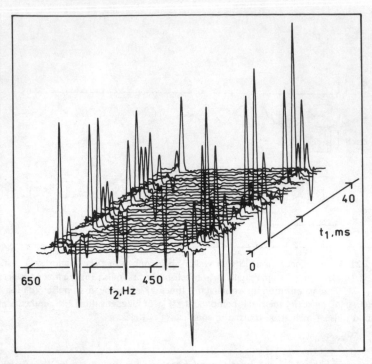

Figure 2.22 Evolution time dependence of the ^{13}C NMR spectrum (one-dimensional) of chloroform, as measured by the pulse sequence of Maudsley and Ernst [29]; stacked-trace plot, phase-sensitive presentation.

periodically in time t_1 with the frequency of the rotation of the component in the x', y' plane [the cosine dependence on the product $t_1 \cdot (f_1 - f_r + J_{IS}/2)$]. As we have learned from an analysis of the SPI experiment above, this type of periodic change in the z' component of magnetization corresponds to a periodic change in the populations of the energy levels. Also, if we measure the spectrum of the S nuclei, the intensities of the lines of the doublet would also vary periodically with the same frequency (i.e., with the frequency of the M^{1+} magnetization rotation in the x', y' plane).

The dependence of the intensities in the spectrum of the S nuclei on time t_1 (the evolution time during which the M^{1+} magnetization was allowed to rotate before the second pulse took place) can be Fourier transformed with respect to t_1 (the second FT in a 2D experiment) to yield the frequency of rotation of the M^{1+} magnetization component during time t_1. The resulting 2D spectrum shows the correlation between the lines in the spectrum of the S nuclei (detected directly) and the frequency of the M^{1+} line in the spectrum of the I nuclei (indirectly detected).

The 2D experiment of Maudsley and Ernst [29] is executed by the pulse sequence shown in Fig. 2.5a. (Actually, the original experiment [29] was more complicated. The roles of the I and S nuclei were reversed to allow sensitive detection of the proton signal. For our understanding of the fundamental aspects of these experiments, the difference is immaterial.) The sequence of two 90° pulses separated by the evolution period (the same sequence as in the Jeener experiment) is accompanied by a "reading" 90° pulse with the frequency of the S nuclei whose response (FID) is detected during time t_2. The first FT of the series of FIDs produces series of spectra of S nuclei $s(t_1, f_2)$ which depend on evolution time t_1. Figure 2.22 shows such a dependence in our example (chloroform) and confirms our expectations.

Since the FID is detected without proton decoupling, according to the pulse sequence in Fig. 2.5a, we see a doublet in each spectrum. Both lines of the ^{13}C doublet change their amplitude periodically (amplitude modulation), and both with the same frequency. While the intensity of one of the lines increases, the intensity of the other line decreases. The frequency of these changes is obtained by the second FT after the spectra $s(t_1, f_2)$ are rearranged into interferograms (remember that an interferogram is the dependence of the signal with frequency f_2 on time t_1). The most significant information is contained in the interferograms that pass through the ^{13}C lines, that is, in sections through the $s(t_1, f_2)$ dependence along the t_1 axis with f_2 equal to the frequencies of the ^{13}C doublets. The result of the second FT of all interferograms, a 2D correlated spectrum, is shown in Fig. 2.23.

Obviously, our expectations are only partially met, since we had predicted a single peak correlating each line of the S doublet with the M^{1+} line. In the spectrum, however, each line of the ^{13}C doublet (a doublet in the direction of the f_2 axis) has associated with it three peaks that have the same f_2 frequency as the ^{13}C line, but with different f_1 frequencies, one of which is zero.

The lack of agreement arises from a simplification in our considerations. We have considered only one line of the I nuclei (to be specific, we have chosen to follow the M^{1+} line). The two 90° pulses, however, affect both lines of the I nuclei, M^{1+} and M^{1-}. The inversions of the two lines (or magnetizations) have opposing effects on the resulting spectra of the S nuclei as we observed when we discussed the SPI experiment. The two components, M^{1+} and M^{1-}, rotate during t_1 with different frequencies (the frequencies of the doublet in the spectrum of nuclei I), and thus the second 90° pulse catches them in different positions. Consequently, their z axis components are also different after the second pulse. The effect on the energy-level populations is a superpositioning of the population changes induced by the z components of the two M^{1+} and M^{1-} magnetizations. The second FT breaks down the observed effect into its frequency components, and that is why each ^{13}C doublet line is correlated with both lines of the ^{1}H doublets. The opposite intensity (opposite intensity sign) of these two f_1 components (or peaks)

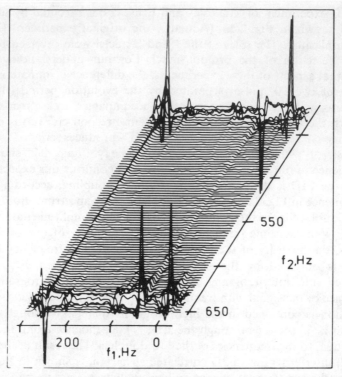

Figure 2.23 Heteronuclear correlated 2D NMR spectrum ($^{13}C-^{1}H$) of chloroform obtained by the second FT of the spectra shown in Fig. 2.22; stacked-trace plot, phase-sensitive presentation to show peaks with negative intensity.

follows from the opposing influence of the two M^I magnetization components on the populations.

We have also neglected the original equilibrium populations. We have seen that these populations bring about intensity asymmetry in the SPI experiment. Since the contribution of the original equilibrium population to the detected signal does not vary periodically with evolution time, the frequency analysis (the second FT) also produces peaks with a frequency f_1 of zero. The original equilibrium population is responsible for the presence of these axial ($f_1 = 0$) peaks in the correlated 2D NMR spectrum.

Summarizing, the positions of cross-peaks with alternating intensity signs correspond to the positions of the lines in 1D spectra of the I and S nuclei. Thanks to spin-spin coupling between the nuclei, which provides the vehicle for polarization transfer, the correlation between spectra of different nuclei can be measured. The coupling creates one common spin system that includes both types of nuclei. Thus information about the motion of one part of the system during the evolution time is transferred to the behavior of the second part of the spin system; we follow this behavior during the detection time. This form of information transfer is sometimes referred to as coherent population mixing [32].

In comparison with other types of 2D measurements, the method described is very sensitive. A polarization transfer from the I nuclei, which have a higher magnetogyric ratio, to the S nuclei, which have a smaller magnetogyric ratio, significantly enhances the signal of the S nuclei. In addition, if the I nuclei relax faster than the S nuclei (protons usually relax faster than carbons), the measurements can be repeated more rapidly (with a short preparation period).

Practical uses of the Maudsley–Ernst pulse sequence are limited, since the 2D spectra produced are complicated by the presence of many peaks, and the measurements require a considerable amount of memory for the data. The actual sensitivity is also somewhat reduced by the presence of axial peaks that do not provide information about correlation of the two spectra. These drawbacks and the means of overcoming them are discussed in detail in Section 4.1.1.

2.5. FUNDAMENTAL 2D EXPERIMENTAL CONSIDERATIONS

The basic idea of 2D spectroscopy should be clear by now, but many questions have no doubt arisen with respect to the practical aspects of performing a 2D experiment. You might ask, for example, how to select the increment Δ of evolution time t_1, how many such increments one should use, what frequency should be selected for the transmitter and decoupler, and so on. Some of these questions are of a fundamental nature, and we address these questions since the operator must understand them when setting up an 2D experiment.

2.5.1. Acquisition Parameters and Memory Requirements

The first problem is the selection of the 2D experiment or pulse sequence, that is, the strategy of using 2D NMR to solve structural problems. The remainder of this book is devoted to this topic. We describe different 2D experiments and discuss the type of problem that can be solved by each experiment. For the moment, let us assume that a pulse sequence has been selected.

The acquisition parameters for the detection period (spectral width along the f_2 axis, sampling frequency, acquisition time t_2, transmitter and receiver frequency, etc.) must be set in the same way as in an 1D experiment (Appendix Section A.6). The parameters related to the evolution period are set in a similar fashion, but we must understand the relationships or correspondence between the parameters of the evolution and the detection periods.

We obtain interferograms after transposition of the spectra produced by the first Fourier transform. Each interferogram consists of the same number of data points as there were Δ increments in the t_1 used in the measurement. In the subsequent (second) Fourier transform the time Δ between two consecutive data points has the same role that the dwell time, t_{dw}, has in the detection period (or in 1D NMR data acquisition). Hence, to represent properly the frequencies along the f_1 axis, Δ must satisfy the condition

$$\Delta \leqslant 1/(2f_1^{max}) \qquad (2.5\text{-}1)$$

where f_1^{max} is the maximum frequency along the f_1 axis (absolute value) in the spectrum.

The value of the maximum frequency f_1^{max} depends on the detection system used. Later in the book we discuss two detection systems (for detailed discussion, see Section A.7), but for the present purpose it should suffice to state that quadrature detection distinguishes the signs of the frequencies (the direction of rotation) and thus allows a positioning of the origin of the frequency axis into the middle of the spectrum. The other detection system (single channel) does not discriminate the frequency signs; therefore, the origin of the frequency axis must be outside the spectrum range.

If the pulse sequence does not permit quadrature detection during the evolution time, the transmitter frequency must be placed at one end of the spectrum (one end of the f_1 axis), and the maximum frequency in the spectrum is equal to the spectral width in the f_1 axis, SW_1 (a situation analogous to single-channel detection during time t_2). Then

$$\Delta^{SD} \leqslant 1/(2SW_1) \qquad (2.5\text{-}2a)$$

When quadrature detection is used, the transmitter frequency is positioned at the midpoint of the spectrum; hence the maximum frequency in the spectrum is only one-half of the spectral width, that is,

$$\Delta^{QD} \leqslant 1/SW_1 \qquad (2.5\text{-}2b)$$

In the detection period it is the receiver reference frequency that determines the origin of the frequency (f_2) axis. (Usually, the receiver reference frequency is the frequency of the observing transmitter, which is called the carrier frequency). The origin of the f_1 frequency axis is determined by the observing transmitter frequency only in homonuclear 2D spectra. In the heteronuclear cases discussed in the preceding sections, the origin of the f_1 axis is determined by the frequency of the decoupler since the decoupler pulses control the evolution during time t_1.

In practice, data economy (see below) dictates that Δ be set to the maximum value allowed by Eq. (2.5-2), but the spectral width should not merely be set equal to the difference between the maximum and minimum f_1 frequencies of the lines in the spectrum. Additional spectral width should be included to incorporate the long tails of spectral peaks with a phase-twisted lineshape.

During a 2D experiment the evolution time is incremented from an initial value of $t_1 = 0$ to the maximum value $t_1^{max} = \Delta \cdot NI$ (NI = number of increments). The maximum evolution time t_1^{max} corresponds to the acquisition or detection time in 1D experiments. Similarly, it determines the achievable resolution along the f_1 axis (see Appendix Section A.6). To have the linewidth LW_1 in the f_1-axis direction, the number of increments must be chosen so that

$$NI \cdot \Delta \geqslant 1/LW_1 \qquad (2.5\text{-}3)$$

(provided that the linewidth given by relaxation time T_2^* is sufficiently less than LW_1).

Let us now examine the economy of data storage. If we need NP (complex) data points in the memory for storing each of our FIDs, we need at least as many (complex) data points for storing each of the spectra obtained after the first Fourier transform. Since we do not want to wipe out the FID data by overwriting spectra into the same memory locations as those used for FID storage (we might later need the FIDs for some other mathematical processing), we need $2 \cdot NI \cdot NP$ complex data points (i.e., $4 \cdot NI \cdot NP$ memory locations) to store these data. In addition, we need memory space to store the 2D spectrum after transposition and the second Fourier transform, and that requires an additional $2 \cdot NI \cdot NP$ memory locations. Obviously, the memory requirements rise sharply with the number of increments used in the experiments, and yet we need a large $NI \cdot \Delta$ in order to have high resolution (narrow lines) according to relationship (2.5-3). Obviously, for the same resolution, the number of increments (NI) can be halved if single-channel detection is replaced by quadrature detection along the f_1 axis (since $\Delta^{QD} = 2 \cdot \Delta^{SD}$). Thus the memory requirements and the measuring time are reduced by a factor of 2 with quadrature detection. (Of course, if measuring time and memory space are not of primary concern, quadrature detection allows us to have double resolution with the same elapsed experimental time and to use the same number of memory locations as in measurements with single-channel detection.)

In every 2D experiment the measurement is repeated several times for each value of t_1. The repetitive data accumulation serves two purposes. First, time averaging of the signal improves the signal-to-noise ratio as in 1D NMR spectroscopy. Second, by varying the relative phases of the pulses and/or the receiver in each scan according to a prescribed phase-cycling scheme, certain features of the signal are canceled and others are constructively added by means of the averaging. The phase-cycling recipes require that the number of scans for the averaged for each value of t_1 must be a multiple of a certain number (often a multiple of 4 or 8 or more). It is therefore important that the number of scans (or transients) to be averaged be set equal to the lowest multiple that will give a sufficient signal-to-noise ratio. Otherwise, spurious peaks may be found in the spectrum if the number of scans is not a multiple of the specified number; also, the experiment may take an exceedingly long time if too high a multiple is used.

We must also pay attention to other aspects of 2D measurements, which, although perhaps not so fundamental, can play almost as important a role in determining the success of the experiment as the fundamental considerations cited above. Some of these factors are discussed in the following sections.

2.5.2. Sensitivity Optimization

Because 2D NMR experiments are time consuming, we must consider optimization of their sensitivity. In a practical sense, each of the 2D experiments to be described requires some optimization, such as finding the best delay or flip angle. There are, however, several general aspects of sensitivity optimization [33, 34], and we shall briefly review several general rules as summarized as "the seven pillars of wisdom" by Levitt et al. [33]:

1. One should estimate spin-lattice relaxation times and the signal envelope $s(t_1, t_2)$ in order to design the optimum experiment. The estimates are usually easy to make after some trial experiments. The decay of the signal along the two time axes clearly shows the meaningful maximum time of t_1 and t_2. Data accumulation after the signal has decayed below the noise level only worsens the signal-to-noise ratio and will not increase resolution more than will filling the data by zeros.

2. It is necessary to choose an optimum length of the preparation period. The delay at the beginning of the preparation period must exceed the spin-lattice relaxation time in order to avoid attenuation of the signal and to reduce noise. This relaxation delay is not always indicated in the pulse

sequences discussed in the following chapters, but it is always tacitly assumed at the beginning of the pulse sequence.

3. One should use the minimum possible resolution along the f_1 axis that is compatible with the purpose of the experiment. That is, use as small t_1^{max} or number of increments NI as possible.

4. If memory space is a problem, use as large a value of Δ as possible. If the data space is unrestricted, small increments of Δ are irrelevant for sensitivity. (The length of Δ does not affect the measuring time needed to achieve the same signal-to-noise ratio.)

5. The detection time, t_2^{max}, and digital resolution along the f_2 axis should be maximized within the limits found in the first step.

6. If the signal envelope (the first step) does not decay monotonically, it might be desirable to incorporate delays into the pulse sequence to provide maximum signal intensity within the ranges of t_1 and t_2 values employed. Examples of such delays are given in the section on correlated spectra with decoupling during detection (Section 4.1.1). With decoupling, one must wait until the various signal components are refocused for the maximum signal.

7. The best sensitivity is achieved by multiplication of the time-domain signal $s(t_1, t_2)$ by a weighting function $h(t_1, t_2)$ that is shaped to match the 2D time-domain envelope of the signal.

2.5.3. Digital Filtering

Very often, the degrading of the resolution by the use of a matched weighting function (the seventh "pillar") cannot be tolerated, and therefore some other weighting function must be used. Multiplication of the time-domain data by a weighting function (sometimes called a window function) has an effect equivalent to an electrical filter in the receiver, but since this filtering is performed digitally, the process is generally known as digital filtering.

A number of mathematical functions have been recommended to serve specific purposes of digital filtering (for a review, see ref. 35). Our choice is limited to the most common functions that are supported by spectrometer software. These weighting functions are known from 1D NMR spectroscopy and are applied to 2D data in the simple form

$$h(t_1, t_2) = h_1(t_1)h_2(t_2) \tag{2.5-4}$$

that is, all the FIDs are multiplied by the weighting function $h_2(t_2)$ prior to the first Fourier transform, and after the transposition the interferograms are multiplied by the function $h_1(t_1)$ before the second transform (the forms of the two weighting functions are not necessarily the same). Since the results of

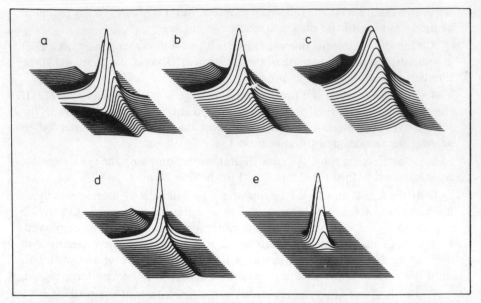

Figure 2.24 2D peaks after simulated digital filtering: (*a*) phase-sensitive presentation of a peak with phase twist, no weighting; (*b*) same peak as in (*a*) but absolute-value presentation; (*c*) matched decreasing exponential weighting, absolute-value presentation; (*d*) increasing exponential weighting, absolute-value presentation; (*e*) resolution enhancement combined with Gaussian apodization, absolute-value presentation. The numerical simulations used the same form of the weighting function in both dimensions. Signal decaying as $\exp(-t_1/2)\exp(-t_2/2)$ was multipled by weighting functions using $a_{-E} = \frac{1}{2}$, $a_{+E} = \frac{1}{4}$, and $a_G = 3$ combined with $a_{+E} = \frac{1}{2}$.

multiplying a FID by these functions are well known from 1D spectroscopy (see e.g., the spectrometer manual), we shall only briefly illustrate their effect on the shape of the most common Lorentzian 2D peak with a phase-twist lineshape. The simulated lineshapes are collected in Fig. 2.24.

For reasons discussed earlier (Section 2.2), a peak with a phase-twist lineshape (Fig. 2.24*a*) is usually presented in an absolute value (Fig. 2.24*b*) or a power spectrum mode. The long dispersion tails and the large linewidth are emphasized if the decreasing exponential weighting function

$$h_{-E} = \exp(-a_{-E}t) \tag{2.5-5}$$

is employed to enhance the sensitivity. The effect of a matched decreasing exponential filter ($a_{-E} = 1/T_2^*$) is apparent from Fig. 2.24*c*.

The convolution difference filter is closely related to function h_{-E} and is described by the function

$$h_{CD} = 1 - A\exp(-at) \qquad (2.5\text{-}6)$$

The convolution difference filter is used to remove a broad portion of a peak from the spectrum with a resultant narrowing of the peak at the base.

When the increasing exponential

$$h_{+E} = \exp(+a_{+E}t) \qquad (2.5\text{-}7)$$

is used as a weighting function, the resolution is increased but the sensitivity is degraded. As is apparent from Fig. 2.24d, the resulting peak is narrow and its contours retain the shape of a four-pointed star.

Line-narrowing by exponential weighting must be used with care to avoid a "clipping" of the signal, which would result in severe line distortions. The distortions can be avoided if the h_{+E} function is combined with some apodization function; most often, Gaussian apodization is used:

$$h_G = \exp(-a_G^2 t^2) \qquad (2.5\text{-}8)$$

The combined weighting function

$$h_{LG} = h_G h_{+E} \qquad (2.5\text{-}9)$$

is known as the Lorentzian-to-Gaussian transform. It gives the best resolution enhancement for a given signal-to-noise ratio [35]. The absorption peaks have circular contours, but in an absolute-value presentation the star shape in general is retained because the dispersive component is not eliminated.

Peak with cylindrical or elliptical contours in an absolute-value presentation can be obtained if the weighted signal decays in a symmetrical fashion on each side of the midpoint of the time axis [36]. Since such signals resemble echo signals, the weighting functions are called pseudo-echo transforms. The most lucid way to obtain a pseudo-echo lineshape is to multiply the decaying signal by the increasing exponential (with $a_{+E} = 1/T_2^*$) in order to cancel the decay of the signal and then to shape the resulting signal so that it decays symmetrically about $t = t^{max}/2$. Using the Gaussian function shifted to the midpoint, the combined weighting function can be written

$$h_{PE} = h_{+E}\exp[-a_G^2(t^{max}/2 - t)^2] \qquad (2.5\text{-}10)$$

The resulting peak is shown in Fig. 2.24e. The lineshape has been transformed from Lorentzian to Gaussian with its circular or elliptical contours and narrow base (the dispersive wings have disappeared). Naturally, the trade-off for this improvement is a significant loss of sensitivity.

Symmetrical shaping of the signal can also be achieved by an h_{GE} filter using $a_G^2 = a_{+E}/t^{max}$ or by employing other symmetrical functions, such as triangular- or sine-wave bell functions.

CHAPTER
3

RESOLVED 2D NMR SPECTRA

A typical (not decoupled) 1D NMR spectrum consists of several multiplets, each of which has a particular value of chemical shift (f, hertz). The separation between the lines of the multiplet is characterized by the value of the spin-spin coupling constant (J, hertz). In cases where the coupling constants are of comparable magnitude to the differences in chemical shifts of the coupled nuclei (i.e., strong coupling), the appearance of the spectrum is complicated by the presence of additional (combination) lines (i.e., a second-order spectrum), and the line separation is then a complicated function of the chemical shifts and coupling constants. Such spectra can be analyzed only by calculations [3, 37, 38] after the lines have been assigned to individual transitions.

A somewhat simpler situation is encountered if the overlapping multiplets belong to nuclei that are not coupled. In such a case we do not have strong coupling that would require computational analysis; nevertheless, it must be decided which line belongs to which multiplet. In addition, some lines belonging to different multiplets may coincide and thus complicate the line assignment problem. There are many possible ways to tackle this problem. Traditional means include decoupling experiments, the use of solvent effects, shift reagents, and so on. The selection of a particular method must be done on the basis of the particular problem, and thus the choice of the method will be dictated not only by the individual problem but also by the experimental means available. The details of the traditional experimental setup must, of course, be chosen carefully, with attention given to all aspects of the problem (availability of the compound under study, its solubility, stability, etc.).

In contrast, 2D NMR spectroscopy offers a universal and systematic solution in a single measurement of one 2D NMR spectrum, and it can be fully automated. The problem can be solved both by "correlated" and by "resolved" 2D NMR spectra. Both types of 2D spectra spread the unresolved and overlapping multiplets of a 1D spectrum according to the NMR parameter in which the overlapping lines differ.

Correlated 2D spectra solve the problem by showing the relationship (correlation) between the lines in an insufficiently resolved spectrum and the resolved lines in another spectrum (e.g., a correlation between unresolved [1]H

NMR lines and resolved ^{13}C NMR lines in the same sample). Resolved 2D spectra solve the problem by separating the effects that the chemical shifts and coupling constants have on the appearance of the spectrum, and they solve the problem without adding any new lines or information to that already present. Depending on the pulse sequence used, the f_1 and f_2 axes spread the spectrum by different combinations of chemical shifts and coupling constants (e.g., in the MKE experiment discussed in Section 2.4.1, the resulting 2D spectra are spread in the f_1 dimension, in which chemical shift and coupling both play a role, while the f_2 coordinate of a peak is determined by chemical shift only). In addition, the separation of the NMR parameters into different axes provides an easy way to determine the values of those parameters (chemical shifts and couplings) that have a definite relationship to the molecular structure.

Resolved 2D NMR spectra can be measured by a number of methods. We begin our discussion with heteronuclear measurements; homonuclear cases are discussed in Section 3.2.

3.1.   HETERONUCLEAR RESOLVED 2D NMR SPECTRA

Attempts to improve the performance of the MKE experiment have produced several variants that can be included under the heading of "gated decoupler methods" [39]. Other methods are based on an insertion of a 180° decoupler pulse into the middle of the evolution period and are known as "proton flip methods" [39], although they are in general not necessarily limited to heteronuclear coupling with protons (thus a more apt name would be "spin-flip methods").

Methods of both groups can produce 2D spectra that are spread according to different parameters. The spreading parameter is usually emphasized in the title of the spectrum. For example, the spectra measured by the MKE method are usually referred to as "chemical shift resolved spectra." We have already shown in Section 2.4.1 that these same spectra can also be viewed as resolved by spin-spin coupling.

The methods of specifying the spreading parameter have not been unified. Some authors refer to all resolved spectra as "J-resolved spectra" or even simply as "J-spectra" (in analogy to the J-spectra measured by spin echo [40]). The latter term is an unsuitable one, since certain correlated spectra are also called J-spectra.

3.1.1.   Gated Decoupler Methods

The simplicity of the basic MKE experiment (Fig. 3.1a) and the suitability of the semiclassical vector model for providing an adequate description of this

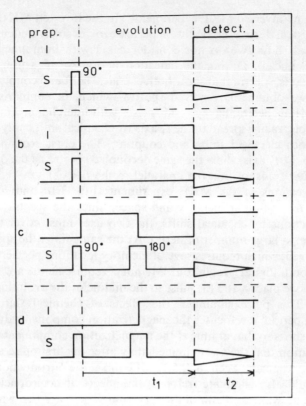

Figure 3.1 Pulse sequences for measurement of heteronuclear resolved 2D NMR spectra by a gated decoupling method: (*a*) MKE experiment; (*b*) inverted MKE experiment; (*c*) decoupling during defocusing; (*d*) decoupling during refocusing. The S nuclei are observed, the I nuclei are decoupled.

experiment (see Section 2.4.1) provide a variety of opportunities for using this experiment to test the basic understanding of the model, the methods, and the concept of 2D spectroscopy.

A simple mathematical operation rotates the MKE spectrum by 90° (transposition). In a rotated spectrum, the stacked-trace plot presents cross sections through the spectrum that are parallel with the f_2 axis. The cross sections that show decoupled spectra are plotted in a left–right direction. 2D spectra with the same appearance can be obtained without the 90° transposition if the pulse sequence shown in Fig. 3.1*b* is used instead of the MKE pulse sequence (Fig. 3.1*a*). This new pulse sequence employs continuous decoupling differently, and while the MKE method uses decoupling only during the detection period, the new sequence uses it only during the evolution period. We therefore call the experiment using the latter pulse sequence an "inverted MKE experiment." In the inverted MKE experiment, the spin system during the evolution period evolves solely under the influence of chemical shifts (we do not consider cases with homonuclear coupling between S nuclei). During the detection period we measure undecoupled spectra, that is, spectra that show both chemical shift and coupling effects. As a result, we obtain 2D resolved spectra, which, in comparison with the original MKE spectra, have f_1 and f_2 axes with interchanged meanings. The inverted spectra are spread in the f_1 axis by chemical shifts only and in the f_2 axis by both chemical shifts and coupling. The cross sections (which are parallel to the f_1 axis) show the same decoupled spectra as the original MKE spectra after transposition (i.e., parallel to the f_2 axis).

Both versions of the MKE experiment (Fig. 3.1a and b) place large demands on computer memory and space. Since the spectra are spread in both directions by chemical shifts, the two axes must cover large spectral widths (i.e., a large number of hertz). At the same time, the spectra must be well resolved, which requires several memory locations per hertz in order to obtain good digital resolution. Memory requirements are considerably reduced if we place a 180°(S) pulse in the middle of the evolution period (i.e., at $t_1/2$). This pulse eliminates the effects of chemical shifts during the evolution period; it refocuses the magnetization components due to different chemical shifts so that at time t_1 the magnetization components are all in the same position that they were immediately after the first pulse at $t_1 = 0$ (see spin echo, Appendix Section A.5.1). Of course, the introduction of a 180°(S) pulse would also eliminate (refocus) the effects of heteronuclear coupling, which we do want to retain in the spectrum. There are two ways to prevent the refocusing of coupling interactions. One method is to apply a 180°(I) decoupler pulse simultaneously with the 180°(S) pulse. The other method is to use continuous decoupling during either the first or the second half of the evolution period [i.e., either before or after the refocusing 180°(S) pulse]. The first method, two simultaneous 180° pulses, forms the basis of the methods with proton (or spin) flip that will be discussed in the subsequent section. Continuous decoupling, during either the defocusing time (Fig. 3.1c) or the refocusing time (Fig. 3.1d), forms the basis of gated decoupling. (The terms "continuous" and "pulsed" decoupling, of course, have only an approximate meaning. For a short time t_1, the duration of a continuous decoupling pulse might be comparable to the duration of a 180° decoupler pulse.) With these modifications, the spin system at the end of the evolution period is in the same state that it would have reached if it had developed only under the influence of spin-spin coupling. The effects of chemical shifts are eliminated, and thus the spectral width in the f_1 axis is significantly reduced, and with it also the requirements on memory data storage and measuring time. A useful side effect of chemical shift refocusing at time t_1 is that the magnetization components due to different chemical shifts are also refocused at $t_2 = 0$; hence the phase correction of the spectra is simpler.

There is no significant difference between methods c and d of Fig. 3.1. Let us look into the mechanism of the latter. In this experiment the M^S, M^{S+}, and M^{S-} magnetization components behave the same after the first pulse as they did in the MKE experiment. That is, in an IS spin system, the sum of the M^{S+} and M^{S-} components, M^S, rotates with a chemical shift frequency $(f_S - f_r)$ in the rotating frame of reference, and the two M^{S+} and M^{S-} components defocus with the frequency of the coupling constant J_{IS}. The refocusing 180°(S) pulse rotates both M^{S+} and M^{S-} components, as in the spin-echo experiment, but as the I spins are decoupled after the refocusing pulse, the angle between the two components no longer changes. The angle (phase) difference acquired during the defocusing period prior to the refocusing pulse is maintained both through the second half of the evolution time (refocusing period) and through the detection period. Thus the phase difference affects the spectrum in the same way as discussed in connection with the basic MKE experiment. It should be noted, however, that the angle between the M^{S+} and M^{S-} magnetization components has been increasing through only half of time t_1, so the angle difference achieved is only half of that achieved in the MKE experiment with the same t_1 time duration. This fact is very important for placement of the correct scale on the f_1 axis, and also when considering the resolution that can be achieved in the f_1 direction. The chemical shifts are refocused by the same mechanism as in spin echo.

The spectra measured by methods c and d of Fig. 3.1 have been spread along the f_1 axis solely by spin-spin interactions (in hertz), and the peaks are separated as the multiplet components. The spreading in the f_2 axis direction is due to chemical shifts only. This means that we have achieved not only a considerable reduction in memory space requirements, but we have also completely separated the two NMR parameters. The f_1 axis shows only the effects of coupling, and the f_2 axis indicates only the effects of chemical shift. (The reader should be able to design similar modifications of the inverted MKE experiment [41].)

An example of a heteronuclear resolved 2D spectrum measured by method d is shown in Fig. 3.2. In comparison with the resolved spectrum measured by the MKE method (Fig. 2.18), the method permits better use of memory space and yields higher digital resolution with the same memory size. Also, a lower number of t_1 increments is needed for the same resolution, and thus measurement time can be saved.

There is another important side effect of the introduction of a 180° pulse or of a spin echo into the sequence for measurement of resolved spectra. The spin echo eliminates the effects of magnetic field inhomogeneity on the evolution during t_1. These effects are eliminated by refocusing at the end of

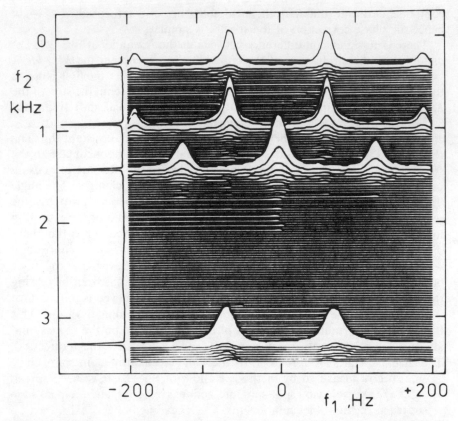

Figure 3.2 Resolved heteronuclear ^{13}C 2D NMR spectrum of 2-butanol measured by pulse sequence of Fig. 3.1d; same sample as in Fig. 2.18; preparation period 5 s; 16 transitions for each t_1 value accumulated, 128 increments of t_1, t_2 = 0.16 s, 400 Hz spectral width in f_1, 3200 Hz in f_2, truncated 256 × 256 data matrix was Fourier transformed and displayed. The FIDs and interferograms were exponentially weighted with a broadening of 3.0 Hz. Total measuring time 3 h. Stacked-trace plot with whitewashing, presentation in absolute-value mode. The 1D decoupled ^{13}C NMR spectrum plotted along the f_2 axis is not a projection; it was obtained in a separate measurement.

the t_1 period (see Appendix Section A.5.1), and therefore the resolution achieved along the f_1 axis is increased; it is limited by molecular diffusion and by spin-spin relaxation. The achievable resolution can be utilized only if we employ a sufficiently large memory for data storage and processing; otherwise, the resolution would be limited by digital resolution. (In other words, the resolution is limited by the frequency interval that is represented by one memory location or by the number of hertz per data point.) The spectrum

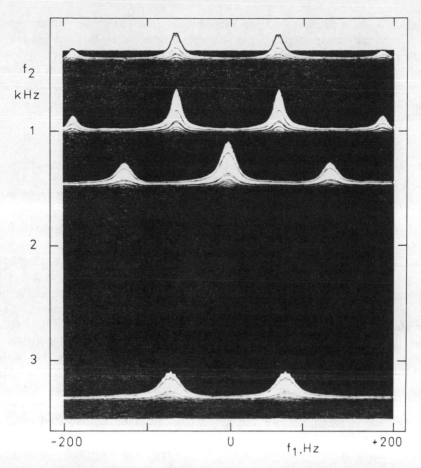

Figure 3.3 Resolved heteronuclear ^{13}C 2D NMR spectrum of 2-butanol measured by pulse sequence from Fig. 3.1d; same experimental data and presentation as in Fig. 3.2 except that the 1024 × 128 experimental FID data matrix was zero-filled to a 2048 × 512 matrix before FT in order to achieve the maximum resolution that the experimental data permit.

shown in Fig. 3.3 demonstrates the increase in resolution achieved simply by using a factor of 16 additional memory locations to process the same experimental data as in Fig. 3.2.

The fact that the peaks remain broad in the direction of the f_1 axis has its origin not only in the absolute-value mode of presentation but is essentially due to many long-range couplings that are small and have remained unresolved. The resolved multiplets correspond to spin-spin coupling between directly bonded nuclei (one bond couplings); couplings with remote nuclei (i.e., with protons two and more bonds apart) are much smaller and there are many such couplings that cannot be resolved by this method (e.g., the carbons of terminal methyl groups coupled with the protons of methylene and methine groups). The methods developed for measurement of spectra that are resolved according to these long-range couplings are discussed in Section 3.1.3.

3.1.2. Spin-Flip Methods

The key to understanding methods based on spin flip is to understand spin-echo, which is described in detail in Appendix Section A.5.1. The methods with spin (proton) flip use a 180° decoupler pulse in the middle of the evolution time to replace continuous decoupling during the evolution period. The simple pulse sequence shown in Fig. 3.4a can serve as a prototype for more elaborate pulse sequences (Fig. 3.4b and c).

The prototype pulse sequence is equivalent to the inverted MKE experiment; note that during t_2 the FIDs of the S nuclei are detected without decoupling the I nuclei (protons). The 180° decoupler pulse in the middle of the evolution period has, at the end of the evolution period, the same effect as continuous decoupling would have had if it had been turned on for the entire duration of the evolution period. The M^{S+} and M^{S-} magnetization components are relabeled by the pulse that converts spin I in the $< + >$ state into the $< - >$ state, and vice versa. Without changing the position of the magnetization components, the relabeling of the I spins has the effect of making the faster M^S component into the slower one, and the slower component is changed into the faster one. For the second half of the evolution time, the former "slow" component, M^{S-}, rotates faster and catches up with the other component, M^{S+}, which was the "fast" component at $t_1/2$, but which now moves more slowly, having been changed by spin I relabeling. The two components refocus at the end of the evolution time, and the spin system is now in a state that does not manifest evolution under the effect of spin-spin interaction. The state depends only on the effects of chemical shifts for the duration of the evolution period because the 180° decoupler pulse does not refocus the chemical shifts of the S nuclei. Hence the 2D spectra are spread in the f_1 axis by chemical shifts only, and in the f_2 axial direction by both chemical shifts and coupling interactions (as in the inverted MKE experiment). Our prototype experiment thus lends itself to improvements analogous to those described for the MKE experiment.

The computer memory is utilized more economically by employing the pulse sequence from Fig. 3.4b. This pulse sequence completely separates chemical shifts (along the f_2 axis) from spin-spin interactions (along the f_1 axis). Following the first 90°(S) pulse, the chemical shift effects during the

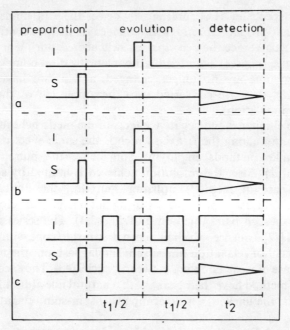

Figure 3.4 Pulse sequences for measurement of heteronuclear resolved 2D NMR spectra using spin flip: (*a*) prototype pulse sequence; (*b*) common version routinely used; (*c*) variant eliminating spin diffusion effects.

evolution period are eliminated by the previously described 180° relabeling pulse on the I nuclei in the middle of the period. This pulse alone would also eliminate heteronuclear couplings, and to retain the information about heteronuclear couplings, the (S) pulse must be accompanied by a simultaneous 180° decoupler pulse on the I nuclei. The decoupler pulse relabels the I spins, and thus it changes the faster component, M^{S+}, into the slower one. The (S) pulse, however, places M^{S+} behind the formerly slower M^{S-} component. The combined effect of the two simultaneous pulses is that the formerly faster component, M^{S+}, now lags behind M^{S-} and becomes the slower M^{S-} component. At the end of the evolution period, it assumes the position which the slower component would have assumed if it had been allowed to develop under the sole influence of coupling interaction for the entire time t_1. Analogously, the formerly slower component, M^{S-}, is turned into the faster component, M^{S+}, and it assumes the position that would have been assumed by this component if it had been left to evolve under a coupling interaction for the entire period and no relabeling took place. Since this pulse sequence provides information on spin-spin interactions along the f_1 axis of a resolved 2D spectrum, it is sufficient to detect the FID under decoupling conditions that produce spectra with chemical shift information only (i.e., along the f_2 axis). Since the decoupler is switched on during acquisition, the angle difference (phases) acquired by the magnetization components during the evolution time are no longer changed, and as we have seen earlier (Section 2.4.1), they are converted into the amplitude of the FID signal (amplitude modulation).

Spin echo eliminates the negative effects of magnetic field inhomogeneity on the resolution along the f_1 axis, which is the same effect as that seen in gated decoupler methods employing spin echo (the pulse sequences of Fig. 3.1c and d). Also, the resolution achieved is limited by spin diffusion, but in the case of methods utilizing spin flip, the limitation can be circumvented, at least in principle, by using not one pair of refocusing pulses but a series of such pairs (method c of Fig. 3.4). The series resembles the Carr–Purcell [42] sequence, which has been designed to cope with spin diffusion in measurement of relaxation times. In a similar fashion, the method works only in the case of weak coupling between I nuclei [43]. The spectra measured by the latter method have their peaks with a natural linewidth in the direction of the f_1 axis; the linewidth is determined by spin-spin relaxation.

3.1.3. Comparison of Experimental Methods

Resolution

Both types of methods for measurement of heteronuclear resolved spectra utilize the features of spin echo during the evolution period. Freeman's group [39] has experimentally confirmed the expectation that the echo effectively eliminates the influence of magnetic field inhomogeneity on resolution along the f_1 axis. The linewidth in the f_1 axis direction is the same in both methods. Methods with spin flip, however, utilize the entire t_1 time for spin system evolution, while the gated decoupler methods utilize only one-half of that time. Hence, when the two methods are used with the same number of t_1 increments, the peak separation and resolution are better in spectra measured with spin-flip methods.

Artifacts

Since methods that employ spin flip use more rf pulses, they are more sensitive to a correct setting of the experimental parameters, especially pulse length.

Pulse imperfections in the observing channel (spin S) can produce artifacts in the 2D resolved spectra measured by either method if the appropriate phase cycling of pulses and receiver is not used. The imperfections (incorrect pulse lengths, inhomogeneity of the pulse field, etc.) give rise to additional lines in 2D resolved spectra, and two types of these additional lines have become known as "phantoms" and "ghosts," respectively. Phantoms are caused by imperfections in both the 90° and 180° pulses, or by an imperfect 180° pulse acting on a system that has appreciably relaxed after the first 90° pulse. The ghosts originate from imperfections in the 180° pulse. Although the phantoms have frequencies in the spectrum that are independent of the measuring method, the positions of the ghosts do depend on the method used for the measurement. Since the magnetic field inhomogeneity is not refocused in these artifact peaks, they can be eliminated by degrading the magnetic field homogeneity [44]. At present, however, all pulse sequences in use employ phase cycling (Exorcycle [45]) that eliminates formation of these artifacts (provided that the correct number of scans for each increment of t_1 was performed in order to complete a full phase cycling).

An imperfection in the decoupler pulses (spin-flip methods) produces an incomplete population inversion, and as a result, the 2D resolved spectrum has altered intensity ratios within its multiplets, and additional lines also appear, the presence of which is in disagreement with the molecular structure. The larger the number of equivalent I nuclei (protons) that interact with the given S nucleus (carbon), the more intense are these additional lines. The lines have frequencies equal to those multiples of $J_{IS}/2$ (hertz) that do not agree with the number of equivalent I nuclei (i.e., do not agree with the multiplicity of the line of the S nuclei) [46]. A typical example is shown in Fig. 3.5, although the spectrum was measured after careful calibration of the decoupler pulse. Pulse inhomogeneity has led to the formation of additional lines; their intensities increase with the number of correct lines in the multiplet, and their frequencies correspond to frequencies in the multiplets with a lower number of correct lines. The formation of these artifacts can be prevented if the simple 180° pulse is replaced by a composite pulse (see Appendix Section A.4.1), but the pulse sequence becomes more complex as a single 180° pulse is replaced by a cluster of three pulses (90°–180°–90°) with different transmitter phases [47].

Strongly Coupled Protons

As long as the I nuclei (protons) are not strongly coupled, the two measuring methods (i.e., gated decoupler and spin-flip methods) are equivalent. When the protons are strongly coupled, the vector model can no longer be used to elucidate the mechanism of the experiment, and quantum physics must be employed to provide reliable results. For this reason we review only briefly the results of such calculations.

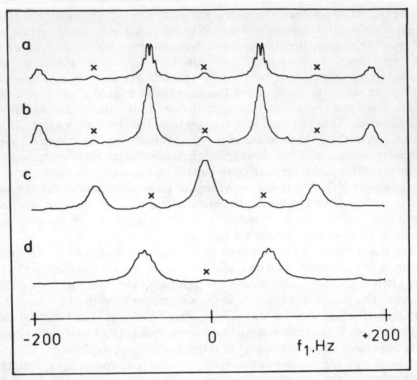

Figure 3.5 Cross section through a resolved heteronuclear (^{13}C-^{1}H) 2D NMR spectrum of 2-butanol measured by the spin-flip method. Sections (*a*) through (*d*) correspond to an increasing f_2 frequency of carbons in CH_3, CH_3, CH_2, and CH groups in the same order as in the spectra in Figs. 3.2 and 3.3; artifacts are marked by "×".

Heteronuclear resolved 2D NMR spectra measured with a gated decoupler have the same structure as conventional 1D spectra, including the asymmetry that is seen in the multiplets of the S nuclei because of strong coupling between protons [48]. Cross sections in such spectra can be displayed in the phase-sensitive mode with a pure absorption lineshape, but the entire 2D spectrum cannot have a pure absorption shape unless it has been measured with the special pulse sequence described by Freeman et al. [49].

The spectra measured by spin-flip methods are symmetrical and consist of multiplets that, on the one hand, do not contain all the lines found in conventional spectra, but on the other hand, do contain additional lines not encountered in conventional 1D spectra. Nevertheless, the spectra can be analyzed on a computer, using suitable programs [50, 51], and exact parameters can then be derived. Some of the lines have negative intensity (and can be overlooked in stacked-trace plots with whitewash).

With some simplification, we can summarize that when the spin system contains strongly coupled protons, it is more convenient to choose gated decoupler methods.

Sensitivity

The pulse sequences described thus far for both types of experiments excite the spin system by applying a nonselective 90° pulse to the system at equilibrium and then detect the FID after the evolution period. During the evolution period the spin system relaxes (thus far we have neglected relaxation in our considerations) and returns to equilibrium. When the t_1 time is long (and relaxation fast), the detected signal is weak. The overall sensitivity per unit time is further reduced by the necessity of waiting during the preparation period until thermal equilibrium is reestablished.

The problem of sensitivity is especially acute in spectra that are resolved according to long-range spin-spin couplings (see "selective measurements" below). These couplings are small, and hence their measurements require use of long t_1 values. In addition, the multiplets are often further split, which means a reduction in line intensity. Obviously, a better signal-to-noise ratio per unit time can be achieved by more efficient excitation and by shortening the preparation period. The preparation period can be shortened by the procedure of Wang Jin-shen et al. [52]. This procedure uses a series of pulses that speed up the return of the spin system to equilibrium.

A more efficient excitation can be achieved by several means. The simple Boltzmann equilibrium distribution that is related to the sample temperature and main magnetic field can be replaced by a distribution that is more favorable for the measurement.

Some of the first measurements of resolved spectra [41, 43] did use decoupling of the I nuclei through the preparation period to enhance the detected signal of the S nuclei. Under the conditions of decoupling, different equilibrium-level populations are established and lead to an enhanced signal of the S nuclei. This "nuclear Overhauser effect" (NOE) [53] can enhance the signal up to a factor of $[\gamma_I/2\gamma_S]$ where γ_I and γ_S represent the magnetogyric ratios of the I and the S nuclei, respectively. In the ^{13}C NMR case, the nuclear Overhauser effect can amount to a twofold signal enhancement when protons are irradiated. The enhancement, however, depends on the mechanism that controls the relaxation of the irradiated nuclei. The maximum enhancement factor above is obtained if the relaxation is dominated by dipole-dipole relaxation. The enhancement achieved on a signal of a particular nucleus S depends on the number of closest I nuclei and on the distance between the S and the I nuclei. Thus the amount of enhancement is small, if any, for quaternary carbons, and it can be close to the maximum value for some methyl carbons. The Overhauser enhancement is easily incorporated into any of the described pulse sequences simply by switching on the decoupler during the preparation period with broadband decoupling of all I nuclei.

The introduction of the other methods of signal enhancement, which are based upon a more effective excitation pulse sequence, are more complicated because they require new programs for pulse sequence generators. The schemes described include excitation by INEPT [54–58] and DEPT [59–60], which are covered in detail in Appendix Section A.5.2. It should suffice here to note that the two pulse sequences utilize polarization transfer from I nuclei (protons, usually) with a high magnetogyric ratio to S nuclei (with a low magnetogyric ratio). The principle of polarization transfer is the same as that in the SPI experiments discussed in Section 2.4.2, but the INEPT and DEPT pulse sequences are arranged in such a way that the transfer does not depend on the chemical shifts of the involved nuclei, i.e., the polarization is transferred irrespective of the chemical shifts of either the S nuclei or the I nuclei. For efficient performance of the sequences, it is necessary to have some estimates of coupling constants J_{IS}. The performance can be optimized for signals of different multiplicities by specifying the pulse timing (INEPT) or the pulse length (DEPT). The enhancement increases with the magnetogyric ratio of the I nuclei and with the number of I nuclei interacting with the S nucleus [58]. Detailed information can be found in the literature. The application of these pulse sequences to resolved spectra has been described by Rutar [61] and by Davis et al. [62].

Selective Measurements

High resolution along the f_1 axis and the separation of signals due to nonequivalent S nuclei along the f_2 axis should allow measurement of small long-range couplings. As is apparent from Figs. 3.3 and 3.5, the splitting due to this coupling is already visible in the spectra measured by the methods described above, provided that sufficient digital resolution can be used and that the presentation is sufficiently detailed. The splitting is apparent as a fine structure of multiplet peaks that are due to much larger coupling between directly bonded nuclei (^1J). Naturally, the splitting is more visible if the spectra are presented in the phase-sensitive mode than in the absolute-value or power spectrum modes.

The presence of small interactions reduces the precision of the determination of large coupling constants, yet small constants cannot be determined with sufficient accuracy from spectra measured by the methods described above. Accurate measurement of small constants requires a large number of FIDs to be measured for different t_1 evolution times, and the maximum evolution time must be sufficiently long to yield narrow lines in the direction of the f_1 axis. The increment Δ in t_1 must be smaller than the reciprocal of the entire multiplet spectral width (in the f_1 axis direction). It therefore follows that it is not possible to determine with sufficient accuracy both small and large coupling constants from only one resolved heteronuclear 2D spectrum. The multiplets with large coupling require small increments; with small increments, however, the maximum t_1 time is limited by the memory size (and overall experiment time), and the large t_1 needed for the measurement of small couplings cannot be attained. Hence it is necessary to modify the general methods for selective measurement of spectra resolved according to small or large coupling constants when both types of coupling interactions are present in the same molecule. Because of the larger variability offered by methods with spin flip, the selective methods have been derived from the pulse sequence shown in Fig. 3.4b.

The simplest approach is to replace the nonselective 180° decoupler pulse by a selective pulse that inverts the spins of a selected group of equivalent I nuclei in the molecule. With a sufficiently selective pulse (a pulse length of about 20 ms), only small long-range coupling effects are apparent in the spectrum [63]. Their small magnitude permits a fine presentation along the f_1 axis, and very precise values can be determined from the cross sections [64]. Naturally, to perform such an experiment, it is necessary to know the resonating frequency for the selective pulse. The selective pulse is most easily realized by setting the transmitter frequency to the center of the multiplet to be selectively irradiated and simultaneously reducing the rf power and adequately prolonging the pulse length in order to maintain the flip angle at 180°. In the resultant spectrum we see only those lines of the S nuclei that are due to spin-spin interactions with selectively inverted I nuclei. One should realize, however, that for these experiments now being considered, it is necessary not to invert the strong center lines of the multiplets of the I nuclei but to invert their satellites that are due to I-S coupling. Since the satellites with small long-range couplings are very close to the center lines (which correspond to no coupling with the S nuclei), very weak and highly selective pulses can assure that the satellites arising from larger one-bond couplings are not refocused by the pulse; in that case, the large couplings will not be apparent in the 2D spectrum.

The other methods are more general and are based on the significant difference in the magnitude of the constants to be measured. These methods [61, 65, 67] replace the 180° decoupler pulse by a $90°_x - d - 90°_x$ [61], or a $90°_x - d - 180°_y - d - 90°_{\pm x}$ cluster of decoupler pulses [65–67]. The latter cluster is an example of a generally useful BIRD cluster that is explained in

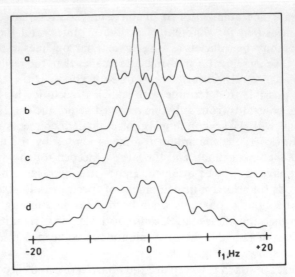

Figure 3.6 Cross sections through a heteronuclear resolved (^{13}C–^1H) 2D NMR spectrum of 2-butanol in hexadeuteriobenzene. Pulse sequence used: $90°(S) - t_1/2 - 90°_x(I) - d - 180°_x(I), 180°(S) - d - 90°_x(I) - t_1/2 - $ detection(S) with $d = 1/(2 \cdot 130)$. The sample shows a doublet of OH protons due to coupling with a CH proton in the ^1H NMR spectrum: analogous spectra of a sample with an OH singlet are much simpler. The cross sections are labeled in agreement with Fig. 3.5.

Section A.5.3. The couplings to be apparent in the resulting 2D spectrum are selected by the choice of the delay, d, and by the relative phases of the pulses (in BIRD). The performance of the pulse sequence of Fig. 3.4b, in which the 180° pulse was replaced by the BIRD cluster to measure small couplings, is demonstrated in the cross sections shown in Fig. 3.6.

The multiplets due to large one-bond couplings have been removed from the 2D spectrum, but the splitting due to small couplings has been retained and is shown with high resolution. The complicated splitting pattern is caused by the large number of small coupling interactions, involving almost all the carbons, by a large number of equivalent protons on a short carbon chain, by coupling with the OH proton, and by strong coupling between the diastereotopic protons of the CH_2 group. Some confusion might be caused by the similar magnitudes of ^3J and ^2J coupling constants, which lead to a virtually similar appearance of cross sections a and b. The apparent quartets are caused by a splitting of triplets (due to coupling with CH_2 protons into doublets because of coupling with the CH proton). Much simpler spectra are obtained for a 2-butanol sample that contains a trace of acid. Then the coupling with the OH proton does not appear in the spectrum. Also, the method described earlier, which uses a selective 180° pulse, produces spectra that are simpler to interpret. Even in the case of our example, however, four 2D spectra would be needed since four proton multiplets should be selectively inverted. In practice, this is not always possible (not only because of excessive time demands but also because the needed selectivity might not be feasible in a crowded spectrum).

Some methods of measurement of correlated spectra also permit selective measurement of coupling constants (see Section 4.1.1).

3.2. HOMONUCLEAR RESOLVED 2D NMR SPECTRA

In contrast to the variety of methods available for the measurement of heteronuclear resolved spectra, there is only one pulse sequence used for homonuclear measurements. It is the spin-echo sequence, with the evolution period covering the time between the first puse and the maximum refocused echo signal. The detection period covers the second half of the echo when the signal is decaying as an FID. Aue et al. proposed using spin echo for this purpose in 1976 [68].

We have already encountered the spin-echo pulse sequence in Section 3.1.2 (and in Section A.5.1). We have seen that to suppress the refocusing of heteronuclear couplings, the 180° pulse in the observing channel must be accompanied by a 180° pulse of the decoupler. In homonuclear systems the single 180° pulse in the observing channel fulfills the two functions of two 180° pulses in heteronuclear systems. On the one hand, it rotates the magnetization components by 180° and thus causes a refocusing of chemical shifts and heteronuclear couplings, and it also eliminates the effect of magnetic field inhomogeneity (as explained in previous sectlons). On the other hand, it also inverts the spins and relabels magnetization components. Consequently, the homonuclear coupling cannot refocus. At the end of the evolution period (i.e., at the instant when the refocusing and the signal are at a maximum), the spin system is in the same state as if it evolved for time t_1 solely under the influence of homonuclear spin-spin coupling; all other effects have been eliminated. The FID is then detected under the same conditions as in conventional 1D measurements.

After the second FT we obtain 2D spectra with peaks whose f_2 coordinates are the same as in 1D spectra (i.e., the line positions are given by the chemical shift and spin-spin couplings). Their position in the direction of the f_1 axis is given by homonuclear couplings only. Obviously, the peaks of a multiplet lie on a line with a slope of unity (i.e., if the same scales are used on both axes, the line is inclined at a 45° angle to the f_2 axis; see Fig. 2.14), and the line intersects the f_2 axis at a frequency corresponding to the chemical

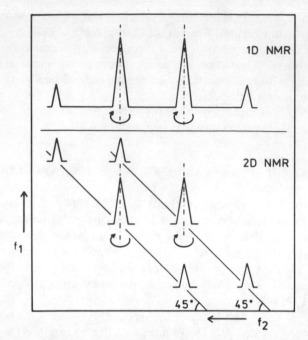

Figure 3.7 Formal "creation" of a resolved homonuclear 2D spectrum from a conventional 1D spectrum, and resolution of an apparent $1:3:3:1$ quartet into two partially overlapping $1:2:1$ triplets. The same scales of Hz/cm are used on both frequency axes; note the direction of the f_2 axis.

shift of the multiplet. You can thus imagine that the 2D spectrum is constructed from a simple 1D spectrum by turning each multiplet 45° around the axis passing through the chemical shift. As demonstrated in Fig. 3.7, this rotation resolves the lines which fortuitously coincide in the 1D spectrum but which belong to different multiplets. Multiplets due to heteronuclear spin-spin coupling are, of course, not rotated.

In practice, the scale used for the f_1 direction usually has much finer gradations (e.g., 0.1 to 2 Hz/cm) than that used along the f_2 axis (e.g., 2 to 10 Hz/cm); while the spectral width in the f_1 direction must be only a few hertz larger than the widest multiplet in the spectrum, the spectral width in the f_2 direction must cover the entire spectrum. For this reason the lines on which the peaks of a multiplet are found are almost perpendicular to the f_2 axis in many spectra (see Fig. 3.8).

Actually, a simple mathematical operation may be invoked to transform the spectrum in such a way that the peaks are on lines that are truly perpendicular to the f_2 axis. The transform (tilting) consists of shifting each

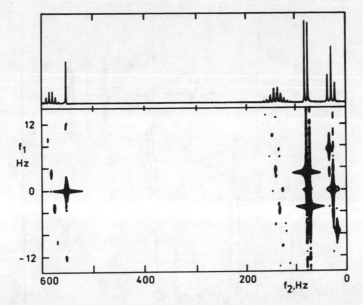

Figure 3.8 Resolved (^{1}H) 2D NMR spectrum of 2-butanol in deuteriochloroform: upper part, conventional ^{1}H NMR spectrum; lower part, contour plot of a 2D spectrum in absolute-value presentation. The multiplet due to nonequivalent diastereotopic CH_2 protons represents a system with strong coupling.

cross section parallel with the f_2 axis by the same number of hertz as is the f_1 coordinate of the cross section.

The tilted spectra (see the example in Fig. 3.9) have completely separated NMR parameters: along the f_1 axis the lines are spread by spin-spin coupling interaction only, and in the direction of the f_2 axis by chemical shifts only.

The peaks in tilted spectra have a distorted lineshape (Fig. 3.9), but the cross sections along the f_1 axis exhibit high resolution for the same reason as for the heteronuclear resolved spectra discussed earlier. Similarly, the effect of spin diffusion can be eliminated by repeating the echo sequence [68]. To utilize the high resolution to the greatest extent possible, a phase-sensitive presentation should be employed, and the experimental parameters as well as the parameters for digital processing of the data after acquisition should also be chosen with care [1, 69].

The tilted spectra can be treated mathematically for noise reduction and removal of excessive "wings" from intense lines [70]. All the important peaks in the tilted spectrum (i.e., those that contain the desired information) are placed symmetrically around axis $f_1 = 0$. (Note that some spectrometer software is capable of producing plots of 2D spectra with positive frequencies

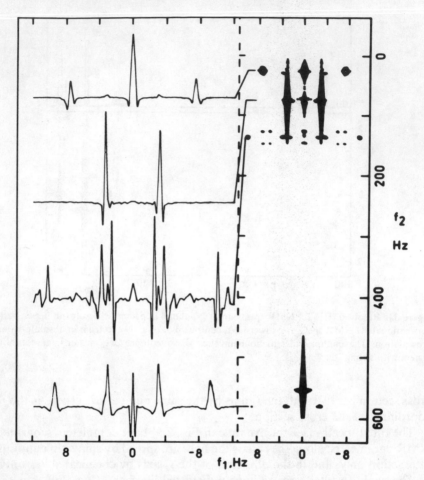

Figure 3.9 Tilted homonuclear resolved ^{1}H 2D NMR spectrum of 2-butanol. The spectrum was obtained by tilting the spectrum shown in Fig. 3.8; the cross sections on the left side are parallel to the f_1 axis, the f_2 frequencies correspond to the positions indicated on the right side of the figure. The sections show high resolution and distorted lineshape at the base because of tilting, the data were resolution-enhanced in the f_1 direction by negative exponential weighting and are presented in the phase-sensitive mode.

only; in such cases the symmetry axis of the tilted resolved spectrum passes through the center of the measured spectral width and is parallel to the f_1 axis.) The wings and noise do not have this symmetry, and thus they can be removed from the spectrum by juxtaposition of signal intensity in the symmetry-related points of the spectrum. In other words, a spectrum with reduced noise and wings can be constructed by replacing the intensity in the

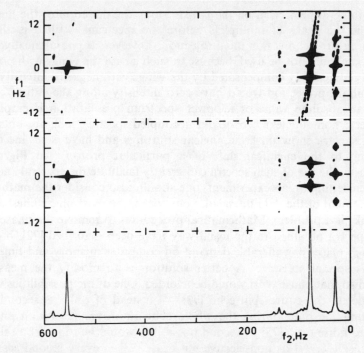

Figure 3.10 Tilted homonuclear resolved ^1H 2D NMR spectrum of 2-butanol and its projection onto the f_2 axis: top, spectrum produced by tilting the spectrum of Fig. 3.8; middle, symmetrized tilted spectrum; bottom, perpendicular integral projection of the tilted spectrum onto the f_2 axis. The strong coupling between CH_2 protons results in the appearance of an additional line in the projection; the line intensity ratios are given by a power spectrum presentation.

two symmetry-related points in the spectrum by the lower of the two intensity values (or by their average). The result of this kind of symmetriz-ation is shown in Fig. 3.10. The procedure is very effective for the measure-ment of large molecules with some strong lines. (For one method of strong singlet-line suppression, see Section 4.2.1 on multiple quantum filtering.)

Projections of resolved spectra are also valuable. A perpendicular integral projection of a phase-sensitive spectrum onto the f_2 axis produces the original 1D spectrum; the same projections onto the f_1 axis yield a J-spectrum [71]. A J-spectrum consists of a superpositioning of all multiplets in the spectrum. The partial projection in which only the spectral region of one multiplet is projected onto the f_1 axis is particularly useful. Such partial projections can be used in the same way as cross sections. The perpendicular projections of phase-sensitive spectra on both axes produce pure absorption line shapes [69] with their inherent advantage of high resolution.

The integral projection on the f_2 axis along the direction of the lines on which lie the peaks of multiplets before the spectrum is tilted is also an interesting projection. For this projection, however, a phase-sensitive presentation should not be used, because in such a case the wings with positive intensity are exactly compensated by the wings with negative intensity, and the resultant projection would have zero intensity along the entire f_2 axis. When an absolute value or a power spectrum presentation is employed, however, "homonuclear broadband decoupled spectra" [68], are obtained; that is, spectra show no homonuclear splittings and have each line corresponding to the chemical shift of a particular proton (see Fig. 3.10). Obviously, the use of such spectra can greatly facilitate the spectral analysis.

When setting up an experiment, one should also consider the mathematical treatment of the data intended. Sometimes, the twist-phase lineshape of the peaks is a problem. Mathematical procedures to remove the phase twist from spectra obtained in the usual way have been described [72], but this approach places considerable demand on computer memory and time, and requires special software. A better solution is to modify the measuring method so that phase twist would be avoided. One of the possibilities would be to detect the entire spin echo [73, 74] instead of only its second half. Collecting the data during the whole echo improves resolution and the signal-to-noise ratio [75]. A second possibility would be to insert a selective 180° pulse [76, 77] or nonselective 90° pulse [78] in every second scan just before the beginning of the detection period.

The resolved spectra of systems with strong spin-spin coupling are more complicated. The effect of a 180° pulse in such systems is much more complex than we have considered; the chemical shifts are not completely refocused. A 2D NMR spectrum contains more lines, some of which may have a negative intensity. The symmetry of the spectrum, as described above, is retained; the peaks are on lines that have the same slope as in weakly coupled systems. The two strongly coupled diastereotopic protons of the CH_2 group in 2-butanol can serve as an example (Fig. 3.8). Spectra of these systems can be simulated and analyzed by computer programs [79, 80] that are generalizations of more familiar programs for the analysis of 1D NMR spectra. It is, however, usually easier to determine the coupling constants or to resolve the spectrum by an indirect measurement of J-spectra that is based on the measurement of heteronuclear correlated spectra (Section 4.1.1).

Practice has shown that it is usually advantageous to combine resolved spectra with other techniques [81, 82].

Some of the overlap that may occur in 2D resolved spectra can be eliminated with multiple quantum filters (MQF), which are considered in Section 4.2.1. Examples of applications of MQF to homonuclear resolved spectra can be found in Ref. 198.

CORRELATED 2D NMR SPECTRA

In measurements of correlated spectra, we utilize polarization or magnetization transfer (population mixing) to pass information about the motion of the spin system during the evolution period on to the detection period. Magnetization can be transferred by different processes. In Section 2.4.2 we have seen magnetization transfer through scalar spin-spin interaction, and we have seen how the populations of energy levels connected by coupling are coherently changed as a result of the action of mixing pulses. Magnetization can also be transferred incoherently through chemical exchange or relaxation processes.

A correlated 2D NMR spectrum indicates which signals are interconnected by magnetization transfer. If we employ spin-spin coupling, the correlation cross-peaks identify the signals that belong to the same spin system; if we use a chemical exchange reaction, the peaks indicate the parts of the molecule between which exchange takes place. We therefore classify correlated spectra according to the mechanism employed for magnetization transfer, and we distinguish spectra correlated through scalar interaction (heteronuclear and homonuclear), through exchange processes, and through double quantum coherence. Other classification schemes can be found in the literature. The individual methods are known by various (and at times rather humorous) acronyms.

Scalar correlated spectra owe their origin to (scalar) spin-spin interaction between nuclei, the same spin-spin coupling that we know from conventional spectra, where it causes splitting into multiplets. Usually, correlated spectra of this type are referred to as "chemical shift correlated spectra." We have already noted that this term is in general not very appropriate. Although a certain type of heteroscalar correlated 2D spectra (decoupled in the directions of both axes) can indeed correlate chemical shifts, the peaks generally show correlation between individual signals. In particular, the name "chemical shift correlated spectra" is misleading in those homonuclear cases in which COSY, SECSY, or FOCSY spectra can be measured in strongly coupled spin systems with a rather complicated relationship between the line positions and the chemical shifts. Since the expression "chemical shift correlated spectra" has been in use for such a long time, there seems to be little hope of replacing it with another more appropriate term, and therefore that title is also used in this book.

2D exchange spectroscopy utilizes an incoherent magnetization transfer. Such transfer occurs when chemical reaction, chemical exchange, or physical exchange between nuclei takes place (e.g., keto-enol tautomerism or hindered rotation), or if there is a dipole-dipole interaction between the nuclei so that the nuclear Overhauser effect (NOE) can be measured.

Because of conceptual novelty, elegance, and power, 2D INADEQUATE spectra are often treated as a third type of 2D spectrum in addition to resolved and correlated spectra. According to our criteria, we include them among correlated spectra since the INADEQUATE pulse sequence includes a mixing period. These spectra correlate chemical shifts in an ordinary 1D spectrum with the frequency of the double quantum coherence.

4.1. CHEMICAL SHIFT CORRELATED SPECTRA

Of all 2D NMR spectra, chemical shift correlated spectra appear to be used most frequently. Their measurement is easy to master, and they can be obtained and interpreted in a routine manner. In addition, the measuring methods are sufficiently sensitive that they do not place unreasonable demands on spectrometer time. Very often, they are applied to problems that could have been solved by conventional means with a little thinking (not necessarily ingenuity) on the part of the experimentalist. The reasons for the popularity of this technique are manifold: in addition to providing the power to solve structural problems, fashionability, ease of automation of the experiment, and the reliability of the method also play a role. The popularity is also due to the qualitative nature of these spectra; they answer in a simple straightforward way our qualitative question about whether two signals are correlated or not. In contrast, 2D exchange spectroscopy must also answer a quantitative question: How fast is the exchange? This question is much more difficult to solve in a reliable and precise way.

We shall now investigate scalar correlated spectra in detail. After scrutinizing simple heteronuclear and homonuclear correlated spectra, which involve one magnetization transfer, we consider RELAY spectra, which include two or more magnetization transfers that occur through either homonuclear or heteronuclear spin-spin interactions.

4.1.1. Heteronuclear Chemical Shift Correlated Spectra (H,X-COSY)

Correlated spectra, which are measured by the procedure explained in Section 2.4.2, contain a wealth of information. In addition to indicating which signals are correlated, they also provide information about the arrangement of energy levels of the spin system. It can be determined from peak intensity signs which of the four energy levels (two for each of the two correlated 1D signals) are progressively arranged and which are regressively arranged [10] (see Section 4.1.2). If, however, we are interested only in line assignment or in similar analytical problems (not an analysis of spin system energy levels), a phase-sensitive presentation is not necessary. In fact, the appearance of peaks with positive and negative intensities does not facilitate interpretation of the spectrum; it is often sufficient to investigate the spectrum in its absolute mode or power mode presentation (provided that degraded resolution is not a hindrance). Even then, the 2D spectrum of such a simple molecule as 2-butanol (Fig. 4.1) is complicated to the extent that interpretation is difficult even when the structure is known.

Guidance for our orientation within the spectrum is provided by the axial peaks (i.e., by those peaks on the axis $f_1 = 0$). These peaks reproduce the conventional 1D ^{13}C NMR spectrum measured without proton decoupling. Accordingly, the signals in the region $f_2 = 550$ to 1050 Hz belong to the methyl group (the axial multiplet in the direction of the f_2 axis is a quartet with a $1:3:3:1$ signal intensity ratio). Signals can be analogously identified in other regions of the spectrum shown in Fig. 4.1. Sections that are parallel to the f_1 axis and pass through the axial peaks represent the corresponding ^{1}H NMR spectra; these sections have the structure of the ^{13}C satellites in ordinary ^{1}H NMR spectra. The doublets (with opposite intensities in a phase-sensitive presentation) are centered around the chemical shift of the proton that is coupled to the given carbon. The separation between the lines of the doublet (in the direction of the f_1 axis) is equal to the direct spin-spin ^{1}H–^{13}C coupling constant. The intensity ratios between the doublets corresponding to different axial peaks of the same multiplet are not the usual ratios of intensities found in multiplets (in contrast to the ratio of axial peak intensities). Although we can calculate these intensity ratios [29, 32], the unusual appearance does not help in interpreting the spectrum.

In the part of the spectrum under consideration (the methyl group, $f_1 = 550$ to 1050 Hz), we find four intense doublets, one doublet for each axial peak. The doublets indicate coupling within the methyl group, that is, coupling between the methyl carbon and the directly bonded protons. The position of the doublets and the separation between the lines of the doublets $[^1J(^{13}C-^1H) = 124.5$ Hz$]$ are in agreement with this assignment; the doublets are centered about the position of the methyl proton quartet in a conventional ^{1}H NMR spectrum. The intensity ratios of the four doublets are $(-1):(-1): 1:1$ (in the phase-sensitive mode) instead of usual $1:3:3:1$ ratios.

In addition to the intense doublets, the cross sections that are parallel with the f_1 axis and pass through the axial peak show some less intense lines. These peaks are due to coupling with the more remote protons, and their position on the f_1 axis corresponds to the shifts of neighboring protons, with the separation between the lines equal to the two-bond or three bond $^{13}C-^1H$ coupling constant (0 to 20 Hz). Peaks of this type (i.e., correlation through two-bond coupling) are also intense in the spectrum of the CH group in Fig. 4.1 (f_2 = 3500 to 4000 Hz region).

If the spectra were plotted on an expanded scale, the cross-peaks would show the same structure as conventional ^1H NMR spectra (e.g., the intense doublets in the spectrum of the methyl group would have a triplet structure, as would the ^1H NMR spectrum of a methyl group attached to a methylene group). In the scale used in Fig. 4.1, these structural features are only roughly suggested for some of the peaks.

It is clear from this description that for practical applications that are not concerned with determination of coupling constants, the measuring method must be modified to yield simpler spectra. The spectra should contain fewer peaks, and the peaks should be well defined (no ambiguity of whether they are due to one-bond or two-bond coupling). It would be desirable to eliminate the axial peaks, since they bring no new information (beyond that which can be obtained from 1D spectra) and reduce the sensitivity of the measurements (because of the dynamic range of spectrometer hardware). For the same reasons, it would be advantageous to use a pulse sequence that permits quadrature detection in the f_1 axis (see Appendix Sections A.2 and A.7). In the case of spectra with excessive multiplet overlap or with sensitivity problems, it would also be desirable to eliminate the effect of $^1H-^1H$ coupling on the correlated 2D spectrum. The means by which these goals can be achieved will be considered in the subsequent paragraphs. It should be noted, however, that the recent progress in NMR instrumentation (high magnetic field, increased sensitivity, and larger computer memory) has brought about a renaissance in interest in 2D correlated spectra, which show all couplings. With performance improved by quadrature detection in the f_1 axis [83], the spectra known as FUCOUP (Fully COUPled) are used in structure determination of complicated compounds [84–87].

Decoupling During the Detection Period

Heteronuclear decoupling (i.e., irradiation of I nuclei while observing S nuclei) is a standard method of spectral simplification in conventional 1D NMR spectroscopy. Spectra measured with decoupling do not show a multiplet structure due to heteronuclear I–S spin-spin coupling; only lines with the frequencies of the chemical shifts are present in the decoupled spectra. Carrying out such measurements is a relatively simple task; the

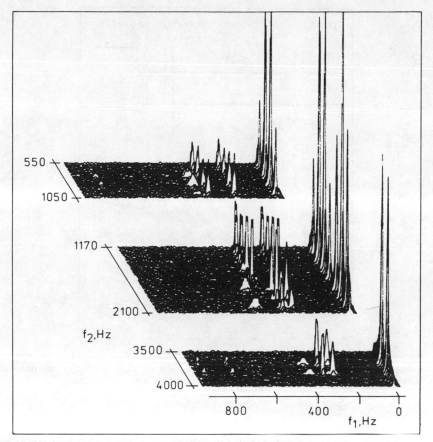

Figure 4.1 Heteronuclear chemical shift correlated ($^{13}C-^{1}H$) 2D NMR spectrum of 2-butanol in hexadeuteriobenzene mesured by the pulse sequence from Fig. 2.5a. Only half of the spectrum with $f_1 > 0$ is shown. Parameters: spectral width in f_1 axis 2100 Hz, in f_2 axis 4500 Hz, quadrature detection in f_2, 512 increments in t_1, 16 scans averaged per increment, FID and spectrum matrix 512 × 4096, preparation period 2 s, total measuring time 5 h. Power-mode whitewashed stacked-trace presentation.

decoupler is either on for the entire duration of the experiment, or it is on through the detection period that immediately follows the excitation (reading) pulse. This simple inclusion of decoupling into the pulse sequence of Fig. 2.5a, however, would not yield informative results.

To explain the situation, let us return to our fundamental heteronuclear correlated experiment that was explained in connection with the pulse sequence of Fig. 2.5a. We know that the two M^{s+} and M^{s-} magnetization components have opposite orientations after the last pulse because of

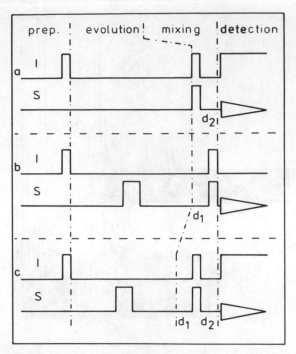

Figure 4.2 Pulse sequences for measurement of heteroscalar correlated 2D NMR spectra: (a) heteronuclear decoupling in f_2 axis; (b) heteronuclear decoupling in f_1 axis; (c) heteronuclear decoupling in both axes. For the length of delays d_1 and d_2, see the text.

polarization transfer. This was demonstrated by the opposite intensities of the doublet lines in the spectra of chloroform in Figs. 2.22 and 2.23. If we were to switch on the decoupler immediately after the last pulse, the two components would maintain their opposite orientations through the entire time that the decoupler is on (i.e., for the entire detection period). In such a case we would detect only a small signal given by the vector sum (the algebraic difference) of the two components. You should recall that in the case of ^{13}C NMR spectroscopy the lines of the doublet have an intensity ratio of $(-3):(+5)$ (polarization transfer from protons). In their vector summation, the part that is due to polarization transfer [the $(-4):(+4)$ doublet] is completely canceled and only the $1:1$ doublet from the equilibrium population is retained (since the components have the same orientation). Thus the experiment with the decoupler switched on immediately after the last pulse would yield the spectra without polarization transfer and the 2D spectrum would contain only axial peaks that are of little interest; however, if we incorporate a short delay, d_2, before we switch on the decoupler (Fig. 4.2a),

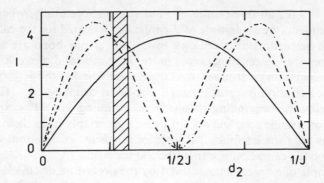

Figure 4.3 d_2 dependence of the signal intensity of S nuclei, as mesured by the pulse sequence of Fig. 4.2a. Only the signal caused by polarization transfer from I nuclei is shown; the solid line is the dependence for the IS system (CH group), the dashed (--) line is the I_2S system (CH$_2$ group), and the broken (-·-) line is the I_3S system (CH$_3$ group). The shaded area around $d_2 = 1/(3J)$ is the favorable region that has an acceptable signal for all three types of spin systems. Intensity $I = 1$ corresponds to polarization transfer without Overhauser enhancement (according to Bax [1]).

the angle between the two M^S components is reduced and the components due to polarization transfer in their vector sum no longer cancel. (This and other delays are often denoted as Δ in the literature. We use d since it is usually used in pulse sequence programming.)

Since the two components rotate with respect to each other with the frequency of coupling constant J_{IS}, the signal would be at a maximum for a delay $d_2 = 1/(2 \cdot J_{IS})$, when the two components are parallel and the contributions from polarization transfer add constructively. During the same delay, the components due to equilibrium populations acquire opposite orientations and cancel each other, and thus no axial peak appears in the 2D spectrum. The 2D spectrum of an IS spin system measured in such a way will therefore contain only two peaks. The f_2 frequency would be the chemical shift of the S nucleus (as in a decoupled spectrum of S), and the two peaks would be centered around the chemical shift of the I nucleus in the direction of the f_1 axis. Their separation would be equal to the spin-spin coupling constant J_{IS}, and the peaks would have opposite intensities in a phase-sensitive presentation.

The optimum delay, d_2, is different for different spin systems, since the M^S magnetization has a different number of components. The optimum delay is that which produces the largest vector sum of these components. The dependence of the signal intensity on the length of the delay d_2 for three common heteronuclear systems is shown in Fig. 4.3.

The choice of $d_2 \approx 1/(3 \cdot J_{IS})$ obviously leads to an acceptably small loss of signal intensity for all three common spin systems. Naturally, when setting up an experiment, the magnitude of J_{IS} must be estimated for the calculation of d_2. Since heteronuclear coupling constants over one bond are markedly different from those stretching across more than one bond, the value chosen for the d_2 calculation determines to a large extent whether the detected signal is due to polarization transfer through one-bond coupling only. (During a short d_2 delay, corresponding to one-bond coupling, the M^S components caused by long-range coupling do not move appreciably from their opposite orientations after the last pulse.) With a longer delay, correlations over two and more bonds can also appear in the 2D spectrum.

An example of a spectrum measured by the procedure discussed above is shown in Fig. 4.4. In comparison with the spectrum shown in Fig. 4.1, the pulse sequence of Fig. 4.2a produces the expected simplification of the spectrum (a decoupled spectrum in the f_2-axis direction).

An optimum d_2 delay is, of course, not the ideal delay, and thus it does not eliminate the axial peaks. Since the delay chosen corresponds to one-bond $^{13}C-^1H$ coupling ($J = 120$ Hz), the cross-peaks from two-bond couplings have been eliminated, and only correlations between directly bonded carbons and protons are indicated. The doublet peaks centered around the proton chemical shift in the f_1 axial direction have remained in the spectrum, and the separation between the peaks in the f_1 direction is equal to the one-bond heteronuclear coupling constant. The peaks have retained the structure of the corresponding multiplets in the ^1H NMR spectra.

Decoupling During the Evolution Period

To simplify further the structure of heteroscalar correlated 2D spectra, we must remove the doublet splitting (doublets in the f_1 direction due to heteronuclear coupling) and the multiplet structure of the peaks in the f_1 direction caused by homonuclear coupling. Removal of each of the two features requires a different modification of the pulse sequence.

(1) Decoupling of Observed Nuclei. Decoupling of observed nuclei during the evolution period should remove the doublet splitting of the peaks by heteronuclear coupling. Although such a simplification is quite desirable, the intention of removing heteronuclear coupling might appear ridiculous since the entire correlation is based on this heteronuclear coupling. The solution offered by Ernst and co-workers [88] is analogous to the procedure used to improve the performance of the MKE experiment (Section 3.1). The suggested procedure combines the insertion of a 180°(S) pulse into the middle of the evolution time with the addition of a delay d_1 at the beginning of the mixing period (the pulse sequence of Fig. 4.2b).

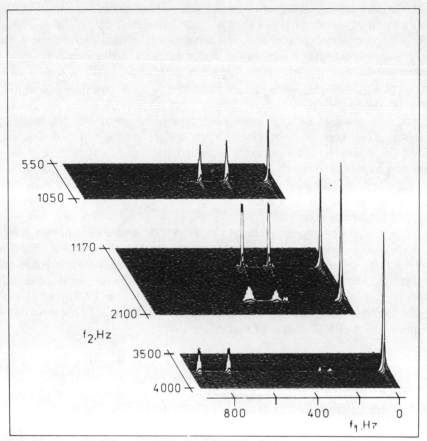

Figure 4.4 Heteronuclear chemical shift correlated (^{13}C–^{1}H) 2D NMR spectrum of 2-butanol measured with decoupling during detection; pulse sequence of Fig. 4.2*a*, delay d_2 = 2.5 ms; sample, acquisition, and presentation parameters the same as in Fig. 4.1.

The 180° pulse interchanges the orientations of the S spins and thus relabels the evolving M^1, M^{1+}, and M^{1-} components of magnetization. The relabeling changes the slow component into the fast one, and vice versa. At the end of evolution period the two components are refocused into the direction of the M^1 magnetization, which corresponds to evolution under the influence of the chemical shift only. During the additional d_1 delay, the two components fan out again. If the delay, however, is chosen so that the components arrive with the opposite orientations, we have the optimum condition for polarization transfer. The ideal d_1 delay is the same for all types of spin systems, (IS, I_2S, and I_3S), $d_1 = 1/(2 \cdot J_{IS})$. The choice of J_{IS} values for calculation of the d_1 delay has a similar role to that discussed above for the calculation of the d_2 delay. With a properly chosen d_1 delay, the M^{1+} and M^{1-} magnetization components will have (prior to the mixing pulses) the opposite orientations (a 180° angle), and their phase will correspond to the evolution for time t_1 under the influence of the chemical shift of the I nuclei.

(2) Homonuclear Decoupling of the I Nuclei (Protons). In many instances, the ^1H–^1H splitting pattern that is apparent in 2D peaks is a welcome feature that helps in assignment or in indirect determination of ^1H–^1H coupling constants (see below). In other instances, the splitting is not desirable and must be eliminated. This occurs when maximum sensitivity is needed (any splitting reduces the sensitivity of the measurement), when the peaks in the 2D spectrum overlap, or when the proton chemical shifts or heteronuclear coupling constants must be determined (splitting reduces the precision of such determinations).

The procedures suggested earlier [66, 89–95] eliminate only homonuclear interactions between protons bonded to different carbon atoms, but the interactions between geminal nonequivalent protons are still apparent in the spectrum. The procedures described use the same principle as the methods for selective measurement of resolved 2D spectra (Section 3.1.3). The 180°(S) pulse for heteronuclear decoupling in the middle of the evolution period is replaced by a BIRD cluster of pulses

$$90^\circ_x(I) - d - 180^\circ(S), 180^\circ_x(I) - d - 90^\circ_{-x}(I)$$

When a delay $d = 1/(2 \cdot ^1J_{IS})$ is chosen, the cluster acts as a 180° pulse on the protons (remote or distant protons) that are not directly bonded to carbon-13, but leaves the protons directly coupled to carbon-13 unchanged. In this way the homonuclear coupling between the directly attached and remote protons is eliminated from the spectrum. (For a detailed discussion of BIRD clusters, see Appendix Section A.5.3).

Total or complete ^1H–^1H decoupling can be achieved by the pulse sequence [95]

$$90^\circ(I) - (d - t_1/2) - 180^\circ(I), 180^\circ(S) - t_1/2$$
$$- 90^\circ(I), 90^\circ(S) - d_2 - \text{decouple}(I), \text{detect}(S)$$

in which the pair of refocusing pulses is incrementally stepped through a fixed evolution period d. (The delay d_2 has the usual role and value.)

The trick of introducing a 180°(I) pulse and keeping the length of evolution period constant works in the following way. The 180° pulse does not affect the evolution under the influence of homonuclear spin-spin coupling (Sections 3.2 and A.5.1). Hence, at the end of the fixed evolution period, the spin system is always influenced by the homonuclear couplings to the same extent, the spectra obtained after the first FT do not show any modulation due to this coupling, and the splittings are not apparent in the resulting 2D spectrum. On the other hand, the proton chemical shifts are refocused by the 180° pulse, but as the pulse is not in the midpoint of the evolution period, the chemical shifts are not refocused at the end of evolution period. The system appears as if it evolved under the influence of proton chemical shifts for a time of duration $(d - t_1)$, and since time t_1 is varied as usual, the spectra are modulated by the proton chemical shifts.

In applications of this method, the length of the fixed evolution time d must be carefully chosen with three factors particularly considered. The chosen value of d should ensure that (1) the proton magnetization components due to heteronuclear coupling are in antiparallel orientations at the end of evolution period (condition for the optimum polarization transfer); (2) at the same time, the components due to homonuclear coupling should not be antiparallel (or they would reduce the extent of polarization transfer); and finally, (3) the evolution time should be long enough to yield sufficiently resolved spectra along the f_1 axis. Obviously, condition 3 cannot be met if the usual value of d_1 [i.e., $1/(2 \cdot {}^1J_{CH})$] is chosen for the value of d. Antiparallel orientation can, however, also be achieved for odd multiplets of this value. The lower multiplets give reasonable polarization transfer over a wide range of ${}^1J_{CH}$ (the recommended value [95] is $d = 0.0105$ s). To avoid polarization transfer loss due to homonuclear coupling, the time should be set $d < (\frac{1}{2} \cdot J_{max})$, where J_{max} is the maximum value of proton-proton coupling in the molecule. The optimum value of d can be found by an INEPT experiment performed for different fixed delays prior to polarization transfer [delay $2 \cdot \tau$ in pulse sequence (A.5-3)]. When high sensitivity is required, homonuclear decoupling can be combined with polarization transfer by techniques such as INEPT [90–92] and DEPT [96].

Decoupling During the Mixing Period

The BIRD cluster of pulses can be also inserted into the middle of the d_2 delay (in, for example, the pulse sequences of Fig. 4.2a or c), when correlation over small interactions is sought. Setting the delay d in the BIRD cluster equal to $1/(2 \cdot {}^1J_{IS})$ and letting the phase of the last decoupler pulse be the same as that of preceding decoupler pulses will have the effect of a 180° pulse on directly attached protons and no effect on the more remote ones. Hence the magnetization components due to direct coupling arrive at the beginning of the detection period with opposite orientations, just as they had after the mixing pulse. The components due to smaller coupling are

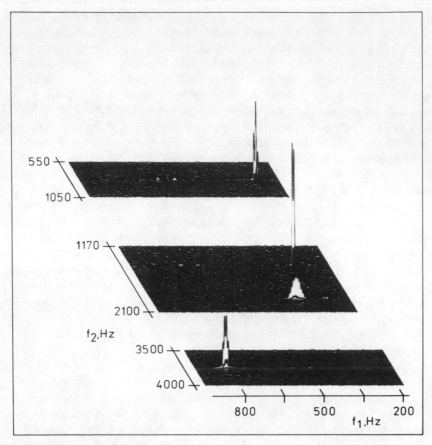

Figure 4.5 Heteronuclear chemical shift correlated (^{13}C–^{1}H) 2D NMR spectrum of 2-butanol measured with heteronuclear decoupling of protons during both detection and evolution periods, quadrature detection in directions of both axes; pulse sequence of Fig. 4.2c, d_2 = 2.5 ms, d_1 = 3.8 ms, FID and spectral data matrices 512 × 2048, spectral width in f_1 axis 1050 Hz, all other parameters as in Figs. 4.1 and 4.4.

refocused, however, provided that the d_2 delay was set sufficiently long (in agreement with the small magnitude of these couplings).

Various forms of decoupling during the evolution period can be combined with decoupling during detection to achieve maximum spectral simplification and increased sensitivity (see the pulse sequence of Fig. 4.2c). The spectrum of 2-butanol measured with combined heteronuclear decoupling during both the t_1 and t_2 periods is shown in Fig. 4.5. The spectrum exhibits only one single peak for each CH$_n$ group, and its position on the f_2 axis corresponds

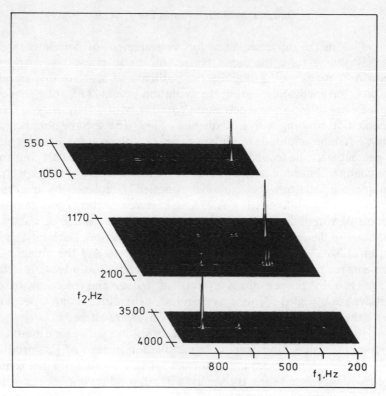

Figure 4.6 Heteronuclear chemical shift (^{13}C–^{1}H) 2D NMR spectrum of 2-butanol measured with heteronuclear and homonuclear proton decoupling; pulse sequence of Fig. 4.2c with the BIRD cluster inserted in the middle of the evolution time, see the text; the same parameters as in Fig. 4.5, except that the number of scans and the total measuring time were reduced by a factor of 2.

with its ^{13}C chemical shift. The peak has the structure and position of the corresponding ^{1}H NMR multiplet in the direction of the f_1 axis. Naturally, with higher resolution (i.e., with a longer evolution time and more memory locations), the structure of the peaks will be better resolved into separate peaks. This 2D NMR spectrum correlates the usual 1D ^{13}C NMR spectrum (measured with proton decoupling) and the corresponding ^{1}H NMR spectrum.

The spectrum shown in Fig. 4.6 demonstrates the additional simplification provided by homonuclear proton decoupling. The decoupling is incomplete for the diastereotopic CH_2 protons, since these protons are both directly bonded to the same carbon atom; therefore, the BIRD cluster does not eliminate the coupling between them.

Quadrature Detection in the f_1 Axis

If all pulses in the pulse sequence for measurement of correlated spectra (Fig. 4.2) always have the same phase (for pulse phases, see Appendix Section A.3), we cannot distinguish the positive and negative frequencies of rotation of magnetization during the evolution period. The pulse sequences have the same effect on magnetization rotating with a frequency $+f$ as on magnetization rotating with a frequency of $-f$. The spectra obtained after the first FT show amplitude modulation; peaks with both frequency signs are then present after the second FT. This is a situation that is analogous to the single-channel detection technique in the measurement of 1D spectra, a technique that has now been almost completely replaced by quadrature detection (see Appendix Section A.2). Quadrature detection distinguishes the direction of rotation of the magnetization, and that brings a number of advantages to the experiment. With quadrature detection, transmitter power is better utilized, memory requirements are reduced, and the number of t_1 increments required is halved. The direction of magnetization rotation can be distinguished by proper phase cycling of pulses and receiver. In each successive scan needed for time averaging for the same length of evolution time t_1, the phases of the pulses and the receiver are altered in a systematic manner. The accumulated FID then produces spectra with phase modulation or a combined phase and amplitude modulation instead of just producing amplitude-modulated spectra. The former types of modulation are sensitive to the frequency sign; hence, the second FT will produce only peaks with the proper sign of the f_1 frequency.

The phase-cycling recipe must accompany the pulse sequence scheme in order to enable successful reproduction of the experiment. The relative phases of pulses and receiver are given as a function of the number of transients (scans or passes), that is, as phases for the nth transient accumulated with the same value of t_1. Phases are usually given in the form of tables, although some pulse programmers require that phases be given as mathematical functions of the number of transients.

Quadrature detection in f_1 permits us to place the decoupler frequency in the middle of the ^{1}H NMR spectrum and to use a weaker field for decoupling with less heating of conductive samples. With the decoupler frequency placed in the middle of the spectrum, the spectral width in the f_1 direction can be half that necessary with single detection. Consequently, the t_1 increment, Δ, can be twice as long. Hence, to achieve the same resolution as with single detection, only half of the number of increments is needed, which is a considerable time saving. With the same computer memory, twice the digital resolution is achieved, or, for the same digital resolution in the f_1 axis, only half of the memory is required. Proper phase cycling also eliminates axial peaks and compensates to some extent for some pulse imperfections.

It is possible to select from two cycling schemes. One of them constructively adds signals from the magnetization components that rotate in both the evolution and the detection periods with the same direction of rotation, and the components that rotate in opposite directions are canceled (antiecho, P-type detection). The other cycling scheme (coherence transfer echo, N-type detection) constructively adds the signals due to the magnetization components that rotate in opposite directions in the two time periods [97–99]. The signal detected with N-type detection has a narrower lineshape. Since the magnetic field inhomogeneities are partially refocused (with different directions of rotation in the two time intervals), the magnetic inhomogeneity components that diverge through the first time interval will converge in the second. This feature can be utilized for considerable line narrowing, as shown by Bolton and Bodenhausen [100].

The drawback of quadrature detection is that it brings with it a twist-phase lineshape. Although individual cross sections through the spectrum with twist-phase lineshapes can be phase corrected, the entire spectrum cannot be corrected. It is therefore necessary to use either the absolute value or the power mode of presentation with a concomitant loss in resolution, especially at the base of the peaks.

When the highest attainable resolution is required (more likely to occur in homonuclear 2D spectra), a more complex approach to quadrature detection must be taken in order to obtain a spectrum with a pure absorption lineshape. Two such approaches have been described in the literature [99, 101, 102], and they can both be applied to heteronuclear correlated spectra. The choice between the method of Marion and Wüthrich [101] and the method of States et al. [102] is dictated by the software of the particular spectrometer. The two methods are discussed and compared in great detail with admirable clarity by Keeler and Neuhaus [99]. In essence, the two methods manipulate data in such a way that the dispersive components of line shape are canceled. Thus a pure absorption line shape is obtained, although at the expense of larger data storage requirements. A more detailed discussion of quadrature detection and pure absorption lineshape can be found in Appendix Section A.7.

The spectrum shown in Fig. 4.5 was measured with simple phase cycling. The pulse sequence of Fig. 4.2c with the phase cycling of Bax and Morris [97] is at present the most commonly used method for measurement of heteronuclear chemical shift correlated 2D NMR spectra. The spectra measured by this pulse sequence are often referred to as HETCOR (HETeronuclear CORrelation) or as H,X-COSY (H-X COrrelated SpectroscopY) spectra. Other methods are used much less frequently. This will probably change as new sequences are provided as a standard part of manufacturer software packages. The fate of a new method is unfortunately affected to a large extent by the ease with which the method is implemented on existing spectrometers.

A practical utilization of heteroscalar correlated spectra is straightforward (although it might be difficult in some complicated molecules). This is perhaps one of the reasons why 2D spectroscopy has become so popular in such a short period of time. Let us see how to work with correlated spectra. We shall again examine the spectrum of 2-butanol. The spectrum shown in Fig. 4.7 is the same as the spectrum in Fig. 4.5, but the stacked-trace plot has been replaced by a contour map presentation.

The usual one-dimensional ^{13}C NMR spectrum of 2-butanol, measured with proton decoupling, is plotted parallel to the f_2 axis (which gives ^{13}C chemical shifts). The ^1H NMR spectrum of the same sample lies along the f_1 axis (the OH line is outside the region shown). The cross-peak in the 2D spectrum connects those lines in the 1D spectra that are due to coupled nuclei. Since we choose d_1 and d_2 delays that correspond to one-bond couplings, the cross-peak connects the carbon line with the multiplet of directly bonded protons. For this reason, spectra such as the one depicted in Fig. 4.7 are called "chemical shift correlation maps" [103]. If we know the assignment of the lines in one of the two 1D correlated spectra (even if only partially), we can assign the lines in the other spectrum through the interconnecting cross-peaks. The procedure is indicated in Fig. 4.7 by the dashed line for the methyl group bonded to the methine group. The multiplet of this group in the ^1H NMR spectrum is identified by its doublet structure due to coupling with the vicinal CH group. As we proceed along the dashed line of Fig. 4.7, the same f_1 coordinate that has the doublet of the methyl group also has the cross-peak that identifies the ^{13}C line of this methyl group.

Obviously, the position of the cross-peak in the 2D spectrum gives the chemical shifts of both nuclei (the value would be determined exactly if we also use homonuclear decoupling). The substituent effects on various nuclei are very often evaluated separately in NMR spectroscopy. Measurement of correlated 2D spectra offers the opportunity to determine the substituent effects as a two-dimensional (vector) quantity that could be useful in analysis of unknown compounds [13].

As in other cases, the textbook example of 2-butanol also points out the weaknesses of the method. If the spectrum contains intense peaks, it is difficult to find all the weak peaks that correspond to the broad multiplets in the ^1H NMR spectrum. If there are only equivalent protons on one carbon atom, this problem can be eliminated by the homonuclear decoupling described earlier. However, we would then lose information about the multiplet structure of the peak, which sometimes can be very helpful in assigning the overlapping ^1H multiplets.

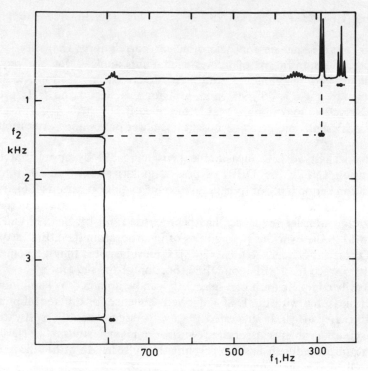

Figure 4.7 Heteronuclear chemical shift correlated (^{13}C–^{1}H) 2D NMR spectrum and 1D ^{1}H and ^{13}C NMR spectra of 2-butanol; identical spectrum as in Fig. 4.5, contour line presentation; the dashed (--) line connects lines due to the C-1 methyl group attached to the methine group.

Selective Measurements

It is often desirable to measure correlations selectively for lines of nuclei having a specified multiplicity (i.e., to have separate correlations for CH, CH$_2$, and CH$_3$ groups) or to measure correlations via coupling constants from a selected narrow range of coupling constant values. (Correlations through small- or long-range couplings are a particularly important class and will be treated separately in the next section.)

Multiplicity selective correlations over a specified range of J values are accomplished by insertion of a polarization transfer portion of INEPT,

$$90^{\circ}_x(I) - \tau - 180^{\circ}(I), 180^{\circ}(S) - \tau - 90^{\circ}_y(I), 90^{\circ}(S) \quad [104, 105]$$

or DEPT,

$$90_x^\circ(I) - 2\tau - 180_x^\circ(I),90^\circ(S) - 2\tau - \theta_y^\circ(I),180^\circ(S) \quad [105, 106]$$

into the pulse sequence for heteronuclear correlation. The selectivity is achieved by optimization of delays τ and flip angle θ for the required multiplicity and coupling constant. As described in Section A.5.2, the optimum flip angle is 45°, 90°, and 135° for CH, CH_2, and CH_3 groups, respectively. The maximum signal is obtained if $\tau = 1/(4 \cdot J_{IS})$. The DEPT sequence is usually preferred because it has fewer pulses and better multiplet selectivity.

The selectivity of 2D measurements is increased by appropriately incorporating INEPT or DEPT types of polarization transfer into the preparation period [105] or (partially) into the evolution period [106], or they can be combined with other procedures [96]. A comparison of several polarization transfer sequences has been carried out by Bendall and Pegg [107], who studied various possibilities of incorporating the DEPT sequence into 2D correlation measurements [108]. It must be noted that it is sometimes difficult to trace the INEPT or DEPT portion of the sequence.

Virtually complete multiplet selectivity can be obtained by combining 2D spectra measured with the DEPT derived sequence for different flip angles. Nakashima et al. [109] suggested a way to circumvent the software and memory size problems associated with adding and subtracting large 2D spectra; the procedure is based on cycling the pulse length within the repeated measurement for the same evolution time t_1.

A different variation of two-dimensional DEPT was proposed by Levitt et al. [110]. The flip angle θ is varied proportionally with the length of evolution time t_1. As a result, the cross-peaks of the CH, CH_2, and CH_3 groups are found according to their multiplicity in different spectral regions on the f_1 axis.

Correlations Through Small Couplings

Despite considerable effort and despite a variety of recommendations and proposed procedures, heteronuclear correlation across several bonds remains a problem even though these correlations could be very important and in many instances decisive in structure determination. If we combine the results of homonuclear correlations and heteronuclear correlations across one bond, we can usually build up the structure of an unknown compound. In case of difficulty, other correlations (e.g., "RELAY" spectra; Section 4.1.3) can help.

If the molecule contains carbon atoms with no directly bonded hydrogens (quaternary carbons), the methods described previously will be of little use. In such a case, it is only possible to determine various fragments of the structure, but we cannot establish by these methods how the fragments are interconnected through the quaternary carbon. One solution is offered by the INADEQUATE method (Section 4.2.2). The method is very elegant, and the spectra can be interpreted in a straightforward manner, but its applications are limited. Since its sensitivity is low, large amounts of sample and measuring time are needed. The other solution would be a heteronuclear correlation across two or more bonds. This correlation would enable us to determine the connectivity of the quaternary carbon by means of its correlations with protons two or more bonds removed (assuming well-defined ranges of magnitude of the corresponding coupling constants, which is not always the case).

The difficulty in measuring these correlations stems from the small values of the coupling constants, which must be considered when setting up the experiment. The heteronuclear and homonuclear coupling constants vary considerably with the structure of the molecule. It should be noted that $^2J(^{13}C-^1H)$, $^3J(^{13}C-^1H)$, and also $J(^1H-^1H)$ vary between 0 and 20 Hz in absolute value, and it is therefore difficult to optimize the delays in the pulse sequences. Nevertheless, many experiments of this type have provided valuable results even though they may have used other than the optimum delays. Of couse, the experiments are time consuming since the values of the coupling constants are small, the delays must be long [the optimum delays are always $1/(nJ)$ seconds], and hence the entire pulse sequence takes an exceedingly long time. During the lengthy execution of the pulse sequence, both the protons and the measured nuclei relax, and we therefore obtain a very weak signal at the end. In such a case it is necessary to accumulate the signal for a large number of transients with the same value of t_1.

COLOC and XCORFE Spectra

The recently proposed COLOC method (COrrelation spectroscopy via LOng range Coupling) [111] attempts to shorten the length of the pulse sequence by incorporating the d_1 delay into the evolution period (the d_1 delay establishes the opposite orientations of the M^{1+} and M^{1-} magnetizations in the pulse sequence from Fig. 4.2c). This reduces the critical time during which a proton relaxes from $t_1 + d_1$ to d_1 only. A pair of simultaneous pulses [180°(I), 180°(S)] is inserted into the d_1 delay, and the evolution time is varied by shifting the pulse pair within the d_1 delay. (See the COLOC pulse sequence in Fig. 4.8a.) Magnetic inhomogeneities are partially refocused, and the protons are decoupled in a homonuclear fashion (as explained in this section in connection with similarly arranged experiments for homonuclear decoupling of protons during the evolution periods and also in Section 4.1.2 in connection with homonuclear experiments with decoupling in the f_1 axis).

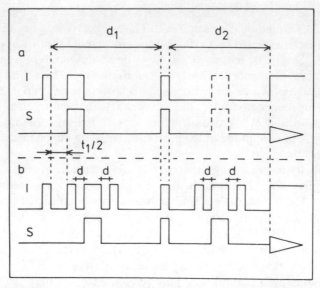

Figure 4.8 Pulse sequences for heteronuclear chemical shift correlations over long-range couplings: (*a*) COLOC; (*b*) XCORFE. The sequences are used almost exclusively with $I = {}^1H$ and $S = {}^{13}C$ for phases in the BIRD cluster (see Section A.5.3); for optimum delays d, d_1, and d_2, see the text.

A maximum suppression of the undesirable signals requires phase cycling with 256 repeated measurements for each value of t_1.

Successful application of COLOC requires experimental optimization of delays d_1 and d_2, because the final effect of the interplay of the various homo- and heteronuclear couplings is difficult to predict. It has been recommended [111] that this optimization be performed via an INEPT experiment. The best refocusing delay d_2 and polarization transfer delay 2τ (Section A.5.2) of the INEPT experiment (performed on an identical sample) should be equated with the d_2 and d_1 delays of the COLOC experiment, respectively. This optimization is easy in the special case of quaternary carbons that are coupled to one group of equivalent protons only (e.g., the carbons of carbonyl groups, as studied in the original work of Kessler et al. [111]). An extremely simple example is provided in Fig. 4.9, the COLOC spectrum of acetone.

Care should be exercised when different nonequivalent protons can be coupled to the same quaternary carbon, because the optimum setting for correlation with one particular remote proton might wipe out correlation peaks of other coupled protons. The method is not suitable for revealing long-range couplings of other than quaternary carbons. As exemplified in

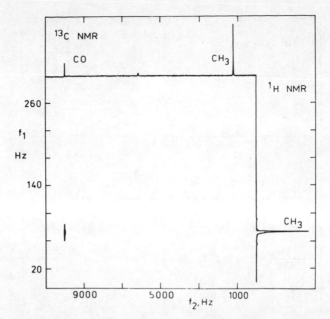

Figure 4.9 (^{13}C–^1H) COLOC spectrum of acetone in hexadeuteriobenzene; 20% solution, pulse sequence from Fig. 4.8a, 256 transients for each of 47 increments in t_1, preparation period 4 s, total measuring time 2 h, optimized delays $d_1 = 0.080$ s and $d_2 = 0.020$ s, spectral widths 11,000 Hz and 300 Hz in f_2 and f_1 axes, 1024 × 256 data matrix for the spectrum.

Fig. 4.10 for 2-butanol, correlations over one-bond couplings are the dominant cross-peaks in COLOC spectra of such carbons, and optimization for long-range polarization transfer is impractical in such cases.

The cross-peaks due to coupling across one bond can be removed if the pairs of 180°(I), 180°(S) pulses in the COLOC pulse sequence are replaced by BIRD pulse clusters (Section A.5.3), as shown in Fig. 4.8b. The new pulse sequence [112] was named XCORFE (X-nucleus-proton CORrelation with Fixed Evolution time) [113]. The introduction of BIRD also brings several other welcome features [112], which are summarized below.

An XCORFE spectrum does not contain correlation peaks over one-bond couplings, while correlation peaks over two bonds show proton-proton coupling, and correlation peaks over three bonds are proton decoupled. These features make the method sensitive and allow differentiation between two- and three-bond correlations, which is otherwise difficult. The proton-proton splittings exhibited on the two-bond correlation peaks are due to three-bond proton-proton coupling involving the correlating proton and the proton bonded to the correlating carbon. (Additionally, as with other

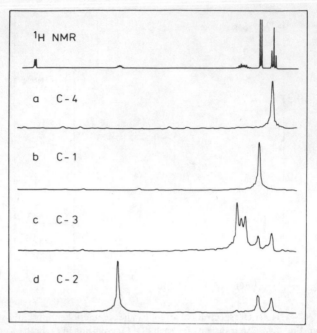

Figure 4.10 (^{13}C–^{1}H) COLOC spectrum of 2-butanol (cross sections along f_1 axis) in hexadeuteriobenzene. Top trace is ^{1}H NMR spectrum, traces a, b, c, and d are cross sections corresponding to chemical shifts of carbons C-4, C-1, C-3, and C-2, respectively; pulse sequence from Fig. 4.8a; experimental parameters the same as in the Fig. 4.5 except for delays d_1 = 0.140 s and d_2 = 0.00551 s; 32 transients accumulated for each of 147 increments in t_1, preparation period 3 s, total measuring time 4 h 15 min.

applications of BIRD clusters, one might expect anomalous behavior when two nonequivalent protons are bonded to the same carbon atom.) Therefore, in an XCORFE spectrum, the line of the carbon marked with the asterisk in the diagram below shows correlation peaks with the protons, also marked with asterisks.

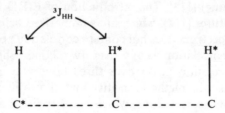

One should realize in evaluating the performance of the XCORFE method that the fixed evolution time d_1 limits resolution along the f_1 axis, yet

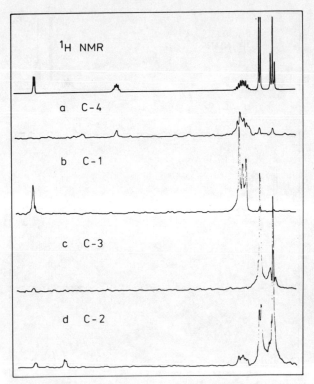

Figure 4.11 (^{13}C–^{1}H) XCORFE spectrum (cross sections) of 2-butanol in hexadeuterioben-zene; pulse sequence from Fig. 4.8b; the same experimental parameters as in the spectrum of Fig. 4.10 except for $d_1 = 0.129$ s, $d_2 = 0.048$ s, and $d = 0.00384$ s; 32 scans for each of 270 increments of t_1, total measuring time 8 h; data exponentially weighted before FT in both dimensions with line broadening of 2.0 Hz; traces shown in the same order as in Fig. 4.10.

d_1 should be chosen as short as possible to avoid signal loss by relaxation. On the other hand, if $d_1 = 1/(2 \cdot ^3 J_{HH})$, polarization transfer would be suppress-ed. Therefore, it is recommended [113] that two XCORFE experiments be performed, one in which d_1 is either 30% larger or smaller than $1/(2 \cdot ^3 J_{HH})$, and the second experiment with a doubled d_1. Examples of two such XCORFE spectra are provided in Figs. 4.11 and 4.12.

Obviously, the two XCORFE spectra are considerably different. The spectrum in Fig. 4.11 contains more features and more information; the experimental settings were closer to the optimum. This is most apparent in trace b (carbon C-1), where the homonuclear coupling suppressed polariza-tion transfer at the end of the short evolution time and thus wiped out the correlation peak with the H-2 protons.

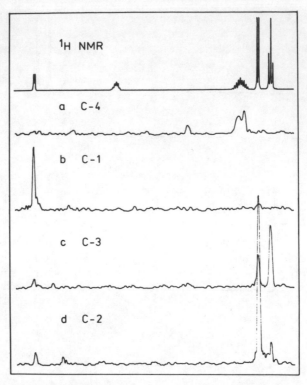

Figure 4.12 (^{13}C–^{1}H) XCORFE spectrum (cross sections) of 2-butanol in hexadeuteriobenzene; same experimental parameters as in Fig. 4.11 except for $d_1 = 0.07$ s and 147 increments in t_1; total measuring time 4 h 11 m.

Despite the helpful structure of the peaks, which will aid in differential correlations over two and three bonds, the XCORFE spectra must be interpreted with considerable care. As illustrated by the example of the spectrum in Fig. 4.11, the traces show the patterns expected on the basis of the discussion above, but some peaks are missing. Thus the C-3 (trace *c*) methine carbon shows a correlation peak with H-1 (singlet) and with H-4 (multiplet) protons, but the expected correlation peak with the H-2 proton is missing. Similarly, the correlation peak of the H-2 proton is missing in cross sections corresponding to the C-1 carbon, and the C-4 carbon shows only a very weak cross-peak with H-2. The conclusion that one might draw from the misleading absence of the H-2 correlation peaks can be remedied by inspection of the C-2 carbon trace, which shows not only a weak correlation peak with the OH proton but also structured peaks with H-3 and H-1 protons and a singlet peak with the H-4 proton.

J Spectra

It has already been mentioned that the proton multiplet structure of the peaks in heteroscalar correlated 2D spectra (measured with heteronuclear decoupling in both dimensions) can be utilized for indirect determination of the homonuclear coupling constants, $J(^1H-^1H)$. The nucleus S whose signal we detect plays the role of a "spy" nucleus by means of which we poke into the proton spin system. Despite the large number of methods available for direct measurement of these coupling constants (both by conventional 1D means and by resolved 2D spectroscopy), the indirect methods receive considerable attention. The indirect methods have two significant advantages over direct ones: first, the overlapping proton multiplets are spread out according to the ^{13}C chemical shifts with large spectral dispersion, and second, proton spin systems with strong coupling are usually converted into systems with weak coupling, which are easier to analyze. While the first advantage should be clear to the reader, the second may require some explanation. The conversion of a strongly coupled system into one with weak coupling is possible because of two well-known features of ^{13}C NMR: the low natural abundance of the ^{13}C nuclei and a much larger magnitude of one-bond couplings than of two- or more bond couplings:

$$^1J(^{13}C-^1H) \gg {}^2J(^{13}C-^1H), {}^3J(^{13}C-^1H)$$

When two strongly coupled protons are bonded to two different carbon atoms, only one of the carbons can be carbon-13 (because of the low natural abundance of ^{13}C). The resonating frequency of the proton directly bonded to the carbon 13 is shifted by $+{}^1J/2$ or by $-{}^1J/2$ (the sign depends on the orientation of the carbon spin) with respect to the frequency of the proton of the same type but bonded to a carbon-12 atom. The adjacent coupled proton also has its frequency shifted, but only by $+{}^2J/2$ or $-{}^2J/2$. The difference in these shifts usually amounts to about 50 to 60 Hz, which is a sufficiently larger difference than the coupling between the two protons, and therefore the multiplet takes on a first-order pattern that is easily analyzed. As in other methods based on the difference between one- and two-bond couplings, the method fails in the case of geminal nonequivalent protons bound to the same carbon atom. The difference in their one-bond coupling constant is too small to convert the spin system to a weakly coupled one, and the multiplets cannot be spread out by the carbon chemical shift.

Minor modifications of the standard pulse sequence of heteroscalar correlated spectroscopy improve its performance for such special purposes in J-spectroscopy as increased resolution in the f_1 axis and multiplet selectivity. The resolution along the f_1 axis is increased by inserting a pair of simultaneous 180° pulses, 180°(I),180°(S), into the middle of the evolution period of the pulse sequence of Fig. 4.2a. The pulses eliminate the effect of proton chemical shifts and refocus magnetic field inhomogeneities [114]. If the splitting due to heteronuclear coupling must also be removed, the pulse sequence of Fig. 4.2c should be used, but the 180°(S) pulse should be replaced by a 180°(I) decoupler pulse [115].

It is advantageous to insert the INEPT [115, 116] or the DEPT [116] cluster of pulses, which would therefore allow editing according to carbon line multiplicity. Multiplets from CH, CH_2, and CH_3 groups can be measured in separate 2D spectra, and thus another possible overlap can be eliminated.

An analogous modification of a pulse sequence can be designed for exact measurements of heteronuclear coupling constants [105, 116], and the relative signs of coupling constants can also be established [117] from a 2D spectrum that is measured by the pulse sequence of Fig. 4.2a in which the flip angle of the second decoupler pulse is shortened by 45°.

Reversed Experiments

As already mentioned, heteronuclear chemical shift correlation experiments, because of the involved polarization transfer, have higher sensitivity than do measurements of resolved spectra. An even higher sensitivity can be achieved by reversed experiments. In these heteronuclear experiments, the much stronger signals of protons (or of other nuclei with high magnetogyric ratios, e.g., ^{31}P or ^{19}F) are detected instead of the weak signals of rare nuclei (which usually have a low magnetogyric ratio, e.g., ^{15}N, ^{13}C) that are detected in conventional heteronuclear experiments. The feasibility and sensitivity of such measurements was first demonstrated by the experiment of Maudsley and Ernst [29], and later by a modified experiment by Bodenhausen and Ruben [118]. The pulse sequence employed by the latter authors contained 10 pulses and caused two transfers of polarization. First, proton polarization was transferred to the rare nuclei (^{15}N), then the spins of rare nuclei evolved for time t_1, and finally, the magnetization was transferred back to the protons and their signal was detected. The resulting 2D spectrum had chemical shifts of rare nuclei along the f_1 axis and proton chemical shifts along the f_2 axis. Similar reversed spectra were recently obtained by combining modern broadband carbon-13 decoupling with the simple one-step polarization transfer sequence of Maudsley and Ernst [29] (modified only to include a refocusing pulse for spectrum simplification [119]).

Although spectrometers on which such reversed experiments can be performed are becoming increasingly available (the spectrometer must permit the excitation and observation of NMR-strong nuclei and the irradiation of rare nuclei as well), the experiments have thus far been used very little because of practical difficulties. The most severe problem is the observation of the signal of protons coupled to rare nuclei in the presence of large signals from all other protons that are not so coupled.

An elegant solution of the technical problems of reversed experiments was suggested by Müller [120] and elaborated by Bax et al. [121–123] and others [124–129]. The solution is based on a heteronuclear multiple quantum coherence (HMQC) that will be explained briefly in Section 4.2. The pulse sequence required can be found in the references cited. (In essence, proton magnetization is transferred initially to zero and double quantum heteronuclear coherences, which are then allowed to evolve. The evolution is interrupted at t_1 and the HMQC is transferred back into proton-observable coherence that is detected; the rare nuclei can be decoupled. Refocusing during the evolution period can be employed to simplify the resulting 2D heteronuclear correlation spectra.) The large sensitivity gain of the proton-detected HMQC experiments (e.g., about 1000-fold for the measurement of ^{15}N chemical shifts [124]) permits correlation experiments and chemical shift measurement of such NMR-difficult nuclei as ^{57}Fe provided that the nucleus with a small magnetogyric ratio has a resolved coupling to a proton (or to another nucleus with a high magnetogyric ratio). The HMQC method can also be adopted for measurement of correlation over long-range couplings [126]. The sensitivity of reversed correlation experiments is such that it permits routine measurements of ^{13}C and ^{15}N chemical shift correlations in proteins at natural abundance levels in an overnight experiment [130].

Heteronuclear multiple quantum filters (Section 4.2.1) and the so-called X-filters [131, 132] also fall under the category of reversed experiments. These devices have been designed to eliminate lines due to protons not coupled to X nuclei from homonuclear 2D spectra.

4.1.2. Homonuclear Chemical Shift Correlated Spectra (COSY)

Homonuclear correlated 2D spectra can, in principle, be measured for any NMR active nuclei. Correlated 2D spectra have even been published for nuclei with quadrupole moments (e.g., ^{11}B), but proton spectra are by far the most frequent. We shall therefore here discontinue the common NMR terminology that uses nuclei designated as I and S and shall speak only about protons in this section. The reader can replace the protons discussed in this section with any favorite nucleus, with the only differences being a lower sensitivity, perhaps a slower relaxation, and larger chemical shifts.

The procedure for measurement of homonuclear correlated spectra, the Jeener experiment [9], has been described in detail in Section 2.1 (Fig. 2.2). The physical mechanism behind this experiment is the same as in the heteronuclear case explained in Section 2.4.2. In the homonuclear experiment, the single mixing pulse fulfills the roles of the two simultaneous (decoupler and observing transmitter) mixing pulses needed in the heteronuclear version.

The first preparatory 90° pulse rotates the equilibrium magnetization of protons into the x, y plane. During the evolution period, various magnetization components rotate with different frequencies. Each has the frequency of the corresponding line in the ^{1}H NMR spectrum, as given by the chemical shift and spin-spin coupling. The second (mixing) pulse transfers magnetizations among various magnetization components. If the transfer takes place between components belonging to the same proton (i.e., transfer within one proton multiplet), the resulting peak is found near the main diagonal in the 2D spectrum. If magnetization is transferred to the component belonging to another nucleus (i.e., a transfer between different proton multiplets), the resulting cross-peak correlates lines belonging to different protons. The cross-peaks can be created only if the two protons are spin-spin coupled. In addition to these magnetization transfers, the second pulse transfers some part of the magnetization into multiple quantum coherence (discussed in Section 4.2.1).

Therefore, the Jeener spectrum represents an autocorrelation diagram of a one-dimensional conventional ^{1}H NMR spectrum. Its use for establishing the relationships between the protons present in the molecule is obvious, and we illustrate it in Fig. 4.13 once again in an example of the spectrum of 2-butanol.

The diagonal (dashed line) runs from the upper left-hand corner to the lower right-hand corner in the contour map. Let us begin with the group of peaks near the diagonal and with the lowest frequencies. These peaks are due to the methyl triplet in the 1D spectrum. The diagonal peaks labeled "A" have the same f_1 coordinate as the AB cross-peak (the other AB cross-peak is placed symmetrically around the diagonal). When we proceed along the dashed line, we move from the AB cross-peak to the second group of diagonal peaks, labeled "B," which belong to protons of the coupled protons of the neighboring CH_2 group (diastereotopic protons). Group B has another and much weaker group of cross-peaks, "BC," which brings us to the diagonal group "C." The C peaks correspond to the one proton of the CH group. This proton is coupled (cross-peaks "CD") with the protons of the terminal methyl group (diagonal peaks "D"). The remaining diagonal peak, which is not accompanied by any cross-peak, is due to the singlet line of the OH proton, which does not show any coupling interaction in the solution being measured. The same procedure is used in analyzing spectra from more complex compounds, or from compounds with an unknown structure. In the procedure described, the crucial point is to find (not to overlook) the weak cross-peaks labeled "BC." These peaks are of very low

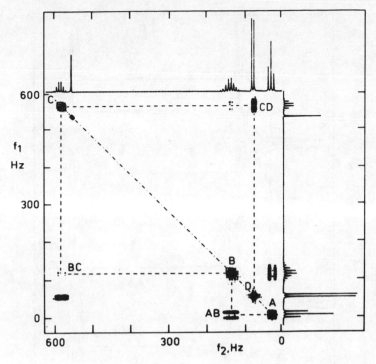

Figure 4.13 Jeener (COSY-90) 2D NMR spectrum of 2-butanol in deuteriochloroform; preparation period 3 s, 512 increments of t_1, FID and spectrum matrices 1024 × 1024, spectral width 680 Hz in both dimensions, four transients for each value of t_1, total measuring time 2.5 h; no digital filtering of the data; the spectrum was symmetrized around the diagonal; absolute-value presentation; quadrature detection in both directions; one-dimensional ¹H spectrum measured in a separate experiment.

intensity since they correlate broad multiplets of low intensities. A more accurate view of the intensities is provided in Fig. 1.1, which presents a stacked trace plot of the lower right-hand portion of the 2D spectrum of Fig. 4.13.

When the Jeener spectrum is presented in the phase-sensitive mode, it has the pattern shown schematically in Fig. 2.15 for the I_3S spin system and in Fig. 4.14 by cross sections through a real spectrum of an IS system. After proper phasing, the peaks near the diagonal have a dispersive lineshape in the directions of both axes. The cross-peaks have a pure absorption lineshape; some of them have a positive intensity, while others have a negative intensity. This pattern is in complete agreement with the analysis given in the classic work of Aue et al. [10].

Although a detailed analysis [10] is beyond the scope of this book, we

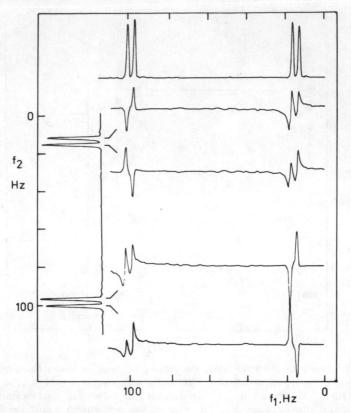

Figure 4.14 Jeener (COSY-90)^{1}H 2D NMR spectrum of 5-bromo-2-furancarboxylic acetamide in hexadeuteriodimethylsulfoxide; cross sections in phase-sensitive presentations. Preparation period 2 s, spectral width in both dimensions 180 Hz, four transients for each of 128 increments, P-type detection, quadrature detection in both dimensions, no digital filtering, FID and spectra matrices 256 × 512.

briefly indicate the most significant results of practical consequence. We use the simplest example, the spectrum of a homonuclear IS spin system with weak spin-spin coupling. The peaks in such a homonuclear correlated spectrum (Fig. 4.15) correlate lines in two identical 1D spectra.

The lines in the 1D spectra are annotated in agreement with the symbolics introduced in Section 2.4 for transitions in the analogous heteronuclear spin systems. The peaks near the diagonal (including those exactly on the diagonal) connect the lines belonging to the same proton; they are peaks of either proton I or of proton S. In accordance with the earlier literature, a pair of lines (or transitions) that produces such a peak is called a "parallel pair"

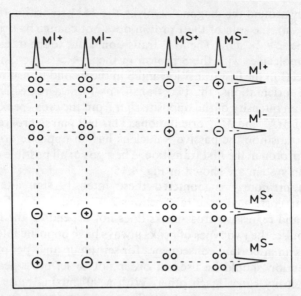

Figure 4.15 Jeener 2D spectrum correlating two identical 1D spectra of an IS spin system. Schematically, the lines in the 1D spectra are labeled as in Section 2.4. The dispersive lineshape of the diagonal peaks is indicated by a "shamrock" (a quartet of circles), and an absorption lineshape is indicated by a circle with a relative intensity sign. Note that the spectrum is rotated by 90° in comparison to the Experimental spectrum shown in Figure 4.14.

[10]. All peaks of parallel pairs of lines (or transitions) have the same phase (i.e., the same proportion of absorption and dispersion lineshapes); the phase depends on the flip angle of the mixing pulse. Those cross-peaks that are farther from the diagonal connect lines that belong to the nonequivalent I and S protons.

The cross-peaks connecting a transition of nucleus I with a transition of nucleus S are classified according to the state of the nuclei that are "passive" in the correlated transitions. (Spin S is passive in the transition of nucleus I, and vice versa.) Either the S nucleus (passive) has the same orientation during the transition of the I nucleus as the I nucleus (passive) has during the transition of the S nucleus, or the passive nuclei have opposite orientations. In the former case, we speak about a regressive pair, and in the latter case, a progressive pair of transitions (or lines). The cross-peaks of progressive pairs have an intensity that is opposite to peaks of the regressive pairs of lines; both have absorption lineshapes.

It is perhaps easier to explain the terms above by using the example of peaks shown in Fig. 4.15. Let us look at one of the two cross-peaks correlating the M^{S+} and M^{I+} lines. The two cross-peaks are placed symmetrically around the diagonal. During the M^{S+} transition (a transition of the S proton), the spin of the I proton does not change its orientation; it remains in the $< + >$ state. The M^{1+} transition is due to the transition of the I proton in molecules with the S proton in the $< + >$ state. Since the two passive nuclei (I in the first transition and S in the second transition) have the same $< + >$ orientations, the two transitions form a regressive pair that should have an intensity of the opposite sign from the cross-peak that is due to the pair of M^{S+} and M^{1-} orientations. This last pair is progressive, since in the M^{1-} transition the passive S nucleus has an opposite orientation to that of the I proton in the first transition. The signs of all possible cross-peaks in the IS spin system are shown in Fig. 4.15.

We have previously encountered these intensity sign alternations in heteronuclear correlated spectra (Sections 2.4.2 and 4.1.1). The phase of progressive and regressive cross-peaks does not depend on the flip angle of the mixing pulse; the two types of peaks always have opposite intensity signs. This fact has an important consequence for setting up an experiment. A high digital resolution should be used in order not to let cross-peaks with an opposite intensity cancel each other. With a 90° mixing pulse (the Jeener experiment), all types of peaks have the same absolute intensity. With a decreasing flip angle of the mixing pulse, however, a decrease is seen in the relative intensities of the peaks correlating transitions that do not share a common energy level. This decrease in intensity is very apparent on all parallel peaks except for those which are exactly on the diagonal. The intensity decrease is the basis for a rapid method of determination of the relative signs of coupling constants [133].

The findings above led to the introduction of a simple modification of the Jeener experiment, one that employs the same pulse sequence but with a shortened mixing pulse. The pulse sequence is

$$90° - t_1 - \alpha - \text{detection} \qquad (4.1-1)$$

These experiments and the resulting spectra are known as COSY (COrrelated SpectroscopY) experiments or spectra; the length of the mixing pulse α is usually specified by indicating the flip angle after a hyphen. For example, COSY-90 denotes the original Jeener experiment with a 90° mixing pulse, and COSY-45 signifies a mixing pulse with a 45° flip angle. Occasionally, when there is danger of confusion, the nuclei are also specified, as in H,H-COSY-45.

A comparison of the COSY-90 and COSY-45 spectra of 2-butanol in Fig. 4.16 illustrates very well the spectral simplification around the diagonal that is obtained with the smaller flip angle. This reduction of the intensity of parallel peaks might be desirable when studying large molecules. Measure-

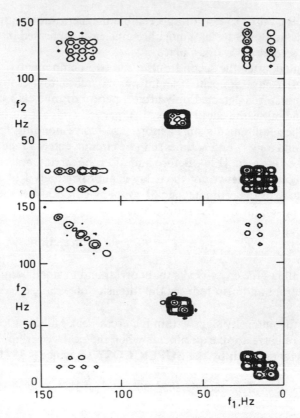

Figure 4.16 COSY-90 and COSY-45 spectra of 2-butanol (detail): top, COSY-90; bottom, COSY-45. With the exception of a shorter mixing pulse for COSY-45, the conditions are the same as in Fig. 4.13.

ment of spectra with a different flip angle provides information necessary for establishing the relationships between spin systems energy levels [10]. The graphical determination of the relative signs of coupling constants is relatively simple and was explained in detailed by Bax and Freeman [133].

The application of COSY spectra for assigning lines to protons in vicinal groups, as described above, is based on the assumption that the cross-peaks do not indicate interactions with coupling constants smaller than vicinal couplings (5 to 20 Hz). At times, however, we do need to establish relationships between more remote protons, which of course are reflected in smaller coupling constants. In order to see the cross-peaks that correspond to these rather small couplings, high resolution in both dimensions must be used (i.e., long t_1 and t_2 times and large data matrices). The magnetization components corresponding to these small couplings fan out very slowly after the first pulse. Also, to get an appreciable magnetization transfer, the two components (in the IS system) must be sufficiently separated in the x, y plane before the second pulse takes place.

Immediately after the second pulse, the two components from magnetization transfer have opposite orientations. A sufficiently strong signal from these components is detected only after a period of time elapses subsequent to the second pulse; the components must partially refocus during this period of time. The requirements on memory size (corresponding to high digital resolution and long t_1 and t_2 times) can be circumvented by the use of a trick suggested by Bax et al. [134]. Before and after the mixing pulse, a fixed delay d is introduced. The length of the delay is approximately $1/(4 \cdot J)$, where J is the magnitude of the estimated small coupling constant between the distant protons. The modified pulse sequence is then

$$90° - t_1 - d - 90° - d - \text{detection}$$

The delay allows the cross-peaks that correspond to small couplings to occur in the spectrum and also reduces the intensity of parallel peaks around the diagonal.

Alternating intensity signs within the cross-peak multiplet lead to reduced sensitivity of correlation experiments when the peaks overlap. This problem can be overcome both by the SUPER COSY sequence [135–138]

$$90° - t_1 - d - 180° - d - 90° - d - 180° - d - \text{detection}$$

(and also by other sequences to be mentioned later). d is a fixed delay between $1/(2 \cdot J)$ and $1/(4 \cdot J)$ which produces spectra with cross-peak multiplets with all peaks having the same intensity sign. Since the diagonal peaks in SUPER COSY spectra have alternating intensity signs, the sensitivity of measurement of the desired cross-peaks is enhanced. Such increased sensitivity is valuable for applications to biological or biochemical compounds with complex and overlapping spectra.

The symmetry of COSY spectra with respect to the diagonal can also be employed for systematic noise reduction. The principle is the same as that described for noise reduction by symmetrization of resolved spectra, but in this case the COSY spectra are made symmetrical around the diagonal instead of around the $f_1 = 0$ axis. An example of the noise reduction achieved is shown in Fig. 4.17.

The COSY spectra (and also other correlated spectra [139]) of complex molecules can be simplified to some extent by decoupling in the f_1 axis [133, 140]. The pulse sequence serving this purpose has a constant measuring time.

$$90° - t_1/2 - 180° - (d - t_1/2) - 45° - \text{detection} \qquad (4.1\text{-}2)$$

The time interval between the 90° and 45° pulses is constant and the refocusing 180° pulse "travels" between the two pulses as the evolution time t_1 is incremented. As we have seen (Sections 3.2 and A.5.1), the 180° pulse refocuses the effects of chemical shifts but does not affect the evolution, which is under the influence of homonuclear spin-spin coupling. Thus the multiplet components, which result from coupling, diverge for the entire duration of the delay d between the 90° and 45° pulses. Since the length of the delay is kept constant for the entire experiment, the coupling influences all spectra measured for different time t_1 values in exactly the same way. As a result, the interferograms are not modulated by couplings, and therefore the peaks in the 2D spectrum will not show splitting along the f_1 axis. On the other hand, although the effects of chemical shifts are refocused at time t_1 after the initial 90° pulse, the magnetizations then evolve under the influence of chemical shifts (only) for the rest of the delay $(d - t_1)$. (The same modification of the evolution period is employed in COLOC and XCORFE experiments although for a different reason; Section 4.1.1.)

The flip angle of the mixing (last) pulse should be small (45°) to avoid mutual cancellation of components (which have opposite phases) within a given cross-peak. Other experimental aspects also have to be carefully considered to obtain the correct results.

Projections of a spectrum measured by this sequence on the f_1 axis yields a ^{1}H NMR spectrum that contains only lines corresponding to proton chemical shifts with no splitting due to spin-spin interactions. In the case of molecules with strong couplings, the spectrum would contain additional lines similar to those found in the projections of tilted resolved spectra (Section 3.2).

Some problems connected with f_1-decoupled COSY spectra can be removed by the method of differential scaling along the f_1 axis. The method employs essentially the same pulse sequence as the f_1-decoupled technique, but the delay $(d - t_1/2)$ is changed to $(1 - k) \cdot (d - t_1/2)$. Then, for a chosen value of k, the apparent chemical shifts are scaled by the factor $(2 - k)$ while the couplings are scaled by the factor k. With a properly chosen value of k, an overlap in an important cross-peak can be removed from the COSY spectrum [141]. Other f_1-scaled experiments are also possible [142].

A very significant improvement in the methods described for measurement of homonuclear correlated spectra (and others as well) has been the introduction of quadrature detection in the directions of both axes. We have already discussed the advantages brought to measurements of heteronuclear correlated spectra by quadrature detection in the f_1 axis. In that case, however, quadrature detection in the f_1 and f_2 axes were independent aspects of the given experiment. Without quadrature detection in the f_1 axis, it was still possible to use simple quadrature detection along the f_2 axis during the detection period (Appendix Sections A.2 and A.7). The two frequency axes are interrelated, however, in homonuclear correlated spectra, and the advantages of quadrature detection cannot be used in one axis without introducing undesirable effects in the direction of the other axis. Thus standard quadrature detection in the f_2 axis can now be utilized in COSY spectra only in those cases where it is also accompanied by quadrature detection in the f_1 axis; quadrature detection in the f_1 axis is accomplished by cycling the phases of pulses and receiver [134]. Quadrature detection should also be combined with phase cycling designed for axial peak suppression, and it can also be combined with CYCLOPS [51] (Section A.3) cycling for elimination of imperfections in quadrature detection in the f_2 axis. Different phase-cycling schemes of quadrature detection are compared in the work of Wider et al. [232], and the handling of data that are required to have spectra with a pure absorption phase is discussed by Keeler and Neuhaus [99] (Section A.7).

As discussed in Section 4.1.1, a choice of two types of quadrature detection exists. With one cycling scheme, we can detect N-peaks, and with the other, P-peaks. Both types of peaks have the same f_2 frequency, but they have different positions on the f_1 axis and are placed symmetrically around the $f_1 = 0$ axis. The N-peaks have a f_1 frequency sign opposite the f_2 frequency; the P-peaks have f_1 and f_2 frequencies of the same sign (the same direction of rotation of magnetization during both the t_1 and t_2 periods; see Fig. A.13). No information will be lost by using the appropriate phase-cycling scheme to select one type of peak only, since peaks of both types contain the same information. Usually, N-type detection is preferred; the N-peaks correspond to a coherence transfer echo (Section A.5.1) with somewhat narrower line shapes than those found with P-type (anti-echo) detection. It is at times important to know which type of detection is incorporated into the pulse sequence being used. The N-type and P-type peaks have an opposite dependence on the flip angle of the mixing pulse, and as already mentioned, this dependence can be very helpful in analyzing the spin system (the arrangement of the energy levels).

With N-type detection it is possible to use the SECSY (Spin Echo Correlated SpectroscopY) modification of the Jeener experiment [143–144]. The modification might be helpful for spectrometers with limited data handling and storing capacity, or when investigating complex or large molecules. COSY spectra of such molecules would contain peaks in an area close to the diagonal, and most of the remaining space would not be used. This situation is remedied by the SECSY variant. The SECSY modification of the COSY pulse sequence delays the beginning of detection by shifting the second (mixing) pulse into the middle of the evolution time; hence the SECSY sequence is

$$90° - t_1/2 - 90° - t_1/2 - \text{detection} \qquad (4.1\text{-}3)$$

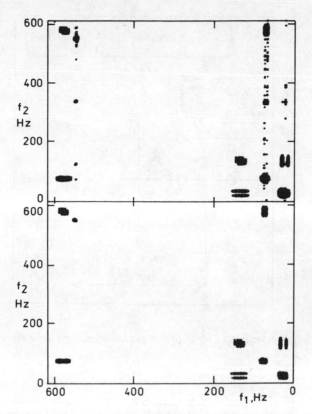

Figure 4.17 COSY-90 spectrum of 2-butanol before and after symmetrization: top, before symmetrization; bottom, after symmetrization, same experimental conditions as in Fig. 4.13.

The magnetization component, which rotates during detection with frequency f_2, has already had this frequency during the second half of the evolution period (after the mixing pulse). During the first half of the evolution period, however, it had a frequency of either $+f_1$ or $-f_1$. By using N-type (echo) detection, we select the component with a $-f_1$ frequency, and hence the average frequency during the entire evolution period is $(f_2 - f_1)/2$. The position of a peak in a 2D SECSY spectrum differs from the position of the same peak in the COSY spectrum, as schematically indicated in Fig. 4.18.

The peaks that correspond to a conventional 1D spectrum are found on a line passing through the center of the SECSY spectrum and parallel to the f_2 axis. The position of the peak in the direction of the f_1 axis is given by the difference of the chemical shifts of the interacting protons, with the peaks being centered midway between the chemical shifts. Cross-peaks lie on lines that are at a 135° angle to the f_2 axis. An evaluation of the advantages and

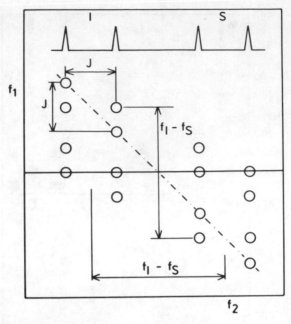

Figure 4.18 SECSY spectrum of an IS proton system (schematic); compare with the COSY spectrum in Fig. 4.15.

disadvantages of COSY versus SECSY is not simple. It is necessary to consider the resolution required in both dimensions (with attention paid both to digital resolution and to line shape, as well as to resolution losses by inhomogeneity in the second half of the evolution time), and also to sensitivity losses because of the delayed detection [1].

For the sake of completeness, the FOCSY (FOldover-corrected Correlated SpectroscopY) method should also be mentioned; it is used primarily for economy of computer time and memory [143]. This variant uses the Jeener pulse sequence, but the t_1 is incremented faster because the increments are larger than 1/(spectral width). With a suitable data arrangement [143], FOCSY spectra that resemble SECSY spectra can be obtained. The scale on the f_1 axis is different, and the cross-peaks are on lines at a 116.6° angle to the f_2 axis instead of at a 135° angle as in SECSY spectra.

Recently, a new class of homonuclear correlation experiments has been introduced. While COSY (or Jeener) spectra contain cross-peaks in the multiplets with alternating intensity signs, the new experiments are arranged so that all cross-peaks have the same intensity sign. The advantages of such experiments were described earlier in this section in connection with SUPER COSY spectra. Physically, the new experiments differ from conventional ones by a different polarization transfer. The conventional experiments are accomplished by a differential magnetization (coherence) transfer; what one line has gained in intensity, the other line in the multiplet has lost (Section 2.4.2), and thus there is no net transfer of magnetization. The new experiments are arranged for a net magnetization transfer. Although several pulse sequences can achieve such a transfer, we shall note only those that seem to have been accepted into 2D NMR practice (detailed discussions of these pulse sequences can be found in the given references, although different terminologies may sometimes be used). All of these pulse sequences eliminate the influence of chemical shifts on the spin system during the mixing period. When the spin system is only under the influence of spin-spin coupling (isotropic), the spins of the system do not behave independently. They take part in one collective motion in which coherences are being transferred throughout the entire coupled spin system in an oscillatory manner (with a period of $1/J$), a process that is generally called isotropic mixing [145–147]. As a result of isotropic mixing, the spectra obtained by the TOCSY (TOtal Correlation SpectroscopY) [145] pulse sequence show cross-peaks between all signals belonging to the same spin system. Therefore, it is possible to identify the whole spin system even if only one proton is separated in the spectrum. Isotropic mixing can also be achieved by cross-polarization under Hartmann-Hahn conditions [148] (which govern the rf field and chemical shifts of the concerned nuclei). HOmonuclear HArtmann-HAhn (HOHAHA) cross-polarization has been obtained by the application of a spin locking [149], by phase-alternated spin locking [149, 150], and by MLEV-17 [151]-based pulse sequences.

Finally, we should also mention a problem that occurs frequently in measurements of homonuclear spectra—the presence of strong lines that bring no useful information. A typical example is the strong singlet line of a solvent (e.g., a water line in biochemical samples). Such strong signals reduce the sensitivity of the method, and their long tails can also mask weak peaks that are important. Many solutions of this problem have been proposed in the literature [152–160]. They range from simple presaturation of the strong line prior to the beginning of the pulse sequence to specially tailored pulses that do not excite the undesirable lines. A very general solution is provided by multiple quantum filters (MQF), as explained in Section 4.2.1. (The analogously named z-filters [161, 162] serve to eliminate phase and intensity distortions and anomalies from the spectra.)

4.1.3. RELAY Spectra

An analytical problem that occurs very frequently is the determination of the sequence of atoms in a molecule, i.e., the determination of which atom is connected to which. NMR methods usually utilize spin-spin couplings to

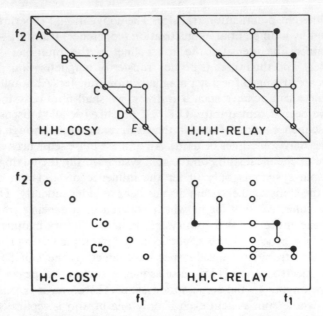

Figure 4.19 Schematic 2D NMR spectra of CH groups A, B, C, D, and E; COSY spectra on the left, RELAY spectra on the right; only the upper half of a symmetric spectrum is shown for homonuclear spectra; the cross-peaks in the RELAY spectra that solve the problem discussed in the text are marked by filled-in circles.

deduce these connectivities in the molecular framework. Prior to the advent of 2D NMR spectroscopy, connectivities were established by interpretation of coupling constants on the basis of comparison with values found in model compounds. 2D NMR spectroscopy offers a much more direct approach, since chemical shift correlated spectroscopy reveals the coupling networks directly.

Difficulties occur, however, (1) if the coupling network is interrupted by a nucleus that has no measurable spin-spin coupling with other nuclei of the network (i.e., a NMR inactive nucleus in the framework), or (2) if the spectra of the two coupling networks overlap. In the former case, the missing link can be overcome by heteronuclear correlation via long-range coupling, which has already been discussed (Section 4.1.1). The latter case, overlapping networks, is often encountered in spectra of oligomers and polymers (e.g., saccharides and peptides) which contain very similar structural units, each of which has an almost identical spectrum. Chemical shift correlated 2D NMR spectra of such species are depicted schematically in the left-hand part of Fig. 4.19.

Obviously, CH groups A, B, C, D, and E belong to two overlapping coupling networks. Because of overlap in the region where group C is found, the Jeener spectrum (H,H-COSY) cannot distinguish whether the two networks are composed of the ACD and BCE groups, or whether the connected groups are ACE and BCD. Heteronuclear correlation (H,C-COSY) shows only that there are two groups in region C, groups C′ and C″, but the connectivity is not established. The problem could also be solved by correlations via long-range coupling or by the INADEQUATE method described in Section 4.2.2. In the present section we demonstrate a different approach that is based on measurement of RELAY spectra.

RELAY spectra solve the problem cited above by means similar to correlations via long-range coupling. RELAY spectra also show correlations between remote nuclei, but it is not necessary to have long-range coupling between remote nuclei for correlation to be revealed in a RELAY spectrum. It is necessary, however, that two remote nuclei (A and X) share a common coupling partner [i.e., a third or intermediary nucleus (M)].

The RELAY pulse sequences are arranged so that magnetization is transferred in the first step from one remote nucleus A to the intermediary nucleus M (via J_{AM} coupling); in the second step magnetization is transferred from M (via J_{MX} coupling) to the second remote nucleus X, the signal of which is then detected. Thus, overall, magnetization is transferred from A to X even though the two nuclei are not spin-spin coupled. With the extent of the first magnetization transfer made dependent on the length of the evolution period, t_1, two-dimensional spectra are obtained.

RELAY spectra (see the right-hand part of Fig. 4.19) usually contain not only RELAY peaks which are due to the two-step magnetization transfer (A-X) but also normal correlation peaks due to spin-spin couplings (A-M and M-X). These two types of peaks in RELAY spectra are easily recognized by comparison with a normal correlation spectrum. In the schematic example that we followed in Fig. 4.19, the RELAY peaks that are marked by filled-in circles prove that the coupling networks are AC′D with BC″E (and not ACE with BCD). A number of real (and more complicated) examples of such problems solved by the use of the RELAY spectra can be found in the literature.

Considering that the three nuclei involved in the simplest relayed magnetization transfer can belong to different nuclear species, many different RELAY spectra and pulse sequences can be anticipated. For example, we can have an all-homonuclear RELAY with magnetization transfer H → H → H (H,H,H-RELAY), or heteronuclear relayed transfers H → X → H (H,X,H-RELAY), or H → H → X (H,H,X-RELAY), and so on. It is also possible to have RELAY of more than two steps [e.g., a four-step homonuclear RELAY (H → H → H → H → H)]. RELAY spectra are sometimes referred to as RECSY (in an analogy to COSY), but more frequently as "relayed COSY" (e.g., H-relayed H,C-COSY).

All RELAY pulse sequences are built from parts of sequences with which we are already familiar, and they are used as building blocks for each of the magnetization transfer steps of RELAY. Thus Bolton [163] used the Jeener sequence (Fig. 2.4a without the detection period) for the first transfer in the first heteronuclear H,H,C-RELAY experiment. The mixing period taken from the heteronuclear chemical shift correlation with decoupling (Fig. 4.2c) then followed. The only modifications of these building blocks are a different d_1 delay and an insertion of two refocusing 180° pulses to ensure the required orientations of magnetization components.

The entire H,H,C-RELAY pulse sequence, which is repeated below, was given in Fig. 2.5c.

(I) $90° - t_1/2 --- t_1/2 - 90°$	$-d_m/2 - 180° - d_m/2 + d_H - 90° - d_2$	decouple
(S) $\qquad\qquad$ 180°	$\qquad\qquad\qquad$ 90°	detect
first polarization transfer	*second polarization transfer*	
evolution period	*mixing period*	

$$(4.1\text{-}4)$$

Let us follow the steps of this pulse sequence. The first two proton pulses transfer magnetization from one proton to the other coupled proton, the amount of transfer depending on the chemical shift of the first proton and also on the J_{HH} coupling, as in the Jeener experiment. The refocusing ^{13}C pulse in the middle of the evolution period eliminates the effects of heteronuclear $^1H-^{13}C$ coupling; the magnetization components due to this coupling are refocused along the directions of magnetization of those protons not coupled to carbons. If we were to detect a proton signal after the second proton pulse, we would observe, as in the Jeener experiment, that some of the lines in the multiplet of the second proton have inverted intensities. (We have discussed the intensity signs of cross-peaks earlier in connection with homonuclear and heteronuclear experiments.) The components of transferred magnetization are antiparallel after the second pulse. Because of the refocusing carbon pulse in the middle of the evolution time, the magnetization components from protons coupled to carbon-13 (^{13}C satellites) are aligned parallel with the corresponding magnetizations from those protons that are not coupled to carbon-13 (i.e., the center lines). The second polarization transfer (proton-carbon) is accomplished by the last pair of 90° pulses [90°(I),90°(S)]. For a successful second transfer step, the ^{13}C satellite components must be antiparallel (Section 2.4.2), and the two components reach an antiparallel orientation after a time delay $d_H = 1/(2 \cdot J_{HC})$ (the d_1 delay in heteronuclear correlations, 3 to 5 ms). For RELAY spectroscopy, however, the second magnetization transfer step should not be selective. It should work equally well for all protons coupled to carbon-13,

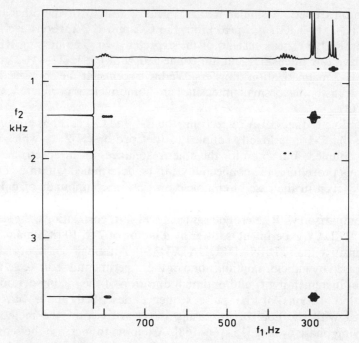

Figure 4.20 H,H,C-RELAY spectrum of 2-butanol in hexadeuteriobenzene; pulse sequence from Fig. 2.5*c*; delays set according to assumed values of coupling constants $J_{HH} = 6$ Hz, $J_{CH} = 130$ Hz, $d_H = 0.00384$ s, $d_m = 0.0378$ s, and $d_2 = 0.00256$ s; all other conditions as given in Fig. 4.7.

and it should not depend on the (second) proton chemical shift or proton-proton coupling constant. Therefore, the antiparallel proton magnetization components must be aligned, which requires an additional time delay $d_m[d_m \approx 1/(4 \cdot J_{HH})$, which will be approximately 30 ms and will ensure that there is no strong selection in favor of a particular proton system multiplicity]. Then the effects of the second proton chemical shift must be eliminated by proton refocusing in the midpoint of this d_m delay. The d_2 delay plays the same role as in heteronuclear chemical shift correlations; thus its optimum value is chosen as described in Section 4.1.1.

An example of the H,H,C-RELAY spectrum of 2-butanol is provided in Fig. 4.20. The spectrum was measured by the RELAY pulse sequence that we have just discussed. The true RELAY peaks can be differentiated from normal correlation peaks by juxtaposition with the heteronuclear correlation spectrum of the same sample shown in Fig. 4.7. Thus we see that the C-4 carbon line (the lowest ^{13}C frequency) is correlated with the H-3 proton multiplet by a true RELAY peak (in addition to normal correlation with the

H-4 proton triplet). A similar RELAY peak is found for the other methyl carbon (C-1, H-2 RELAY peak), but the C-2 and C-3 carbons each show only one RELAY peak instead of the expected two. The missing RELAY peaks have low intensities and are connected with the H-2 → H-3 and the H-3 → H-2 magnetization transfers. We have seen that the cross-peaks due to this transfer have small intensities in homonuclear correlated spectra (Fig. 4.13).

According to the RELAY spectrum, the C-1 and C-2 carbons are adjacent because the C-1 is indirectly coupled to that proton (H-2) to which C-2 is directly coupled, and so on for the other resonances in the spectrum. By an analogous procedure, the connectivity can be determined within a coupling network (even in the case of complete overlap in compounds of unknown structure).

In comparison with heteronuclear chemical shift correlations, the sensitivity of a RELAY experiment is lower by a factor of 2 to 10 (it remains much more sensitive than 2D INADEQUATE) [164]. Although sensitivity can also be increased by judicious optimization of the experimental setup (e.g., proper choice of incrementing t_1 and optimum duration of the mixing period [165], some modifications of the pulse sequence described above have been proposed with the goal of increasing the sensitivity of the method and simplifying phase cycling [15, 166–168]. Attempts to increase the sensitivity also include heteronuclear RELAY with proton detection (reversed experiments) [129, 169–172] and Hartmann–Hahn-type cross-polarization [150, 172]. (The latter, HAHA experiments, although very promising, are beyond the scope of this book. A description of these experiments can be found in the work of Bax et al. [150, 172].) Other modifications address the basic problem of RELAY spectroscopy: discrimination between true RELAY peaks (often called "remote" cross-peaks) and normal correlation peaks ("neighbor" peaks). The undesirable "neighbors" have been suppressed by using BIRD clusters [173], or low-pass J-filters [174], or by using proton double-quantum coherence as the correlation frequency [175].

When optimization of the mixing delay ($d_m + d_H$) for the second transfer is difficult, the delay can be incremented proportionally to the evolution time t_1 throughout the experiment in a manner analogous to that used in "accordion" exchange spectroscopy (Section 4.3.2) [176]. This approach is especially convenient in multistep RELAY experiments [177].

The homonuclear version of the RELAY experiment, H,H,H-RELAY, uses a $[-\tau - 180° - \tau - 90° -]$ sequence for the second (mixing) magnetization transfer step, which can be repeated for multistage RELAYs [177–179]. Thus the simplest two-step RELAY pulse sequence is

$$90° - t_1 - 90° - \tau - 180° - \tau - 90° - \text{detect}$$

(Phase cycling should eliminate undesirable contributions from the nuclear Overhauser effect [179, 180]). Practical guidelines for optimization of experimental parameters of homonuclear RELAY have been given by Bax and Drobny [179, 181].

Although it might appear to be a curiosity, H,X,H-RELAY pulse sequences have also been designed and used as an extreme means of determining the structure of a compound with overlap in the spectrum of the heteroatom [182] or with two completely isolated coupling networks [183]. Measurement of H,C,C-RELAY spectra have low sensitivity, although they still have a higher sensitivity than INADEQUATE [164].

4.2. SPECTRA CORRELATED BY MULTIPLE QUANTUM COHERENCE

In this section we introduce the principles and uses of some of the most elegant and powerful methods of 2D NMR. The practical aspects of the uses of these methods are relatively easy to convey, but the principles are more difficult to explain at our level of treatment. There are two causes of this difficulty. First, the methods are based on the concept of multiple quantum coherence (or transitions), which is not generally familiar, and second, multiple quantum coherence cannot be adequately treated by the vector model that we have been using. Therefore, we must limit our explanation to a description of the results obtained by rigorous quantum mechanical methods (which employ density matrix formulations). A reader who would like to learn more about the interesting phenomenon of multiple quantum coherence (or spectroscopy) is referred to the excellent review by Bodenhausen [184] and a more general recent study [2, 185].

4.2.1. Multiple Quantum Coherence, Coherence Transfer Pathways, and Multiple Quantum Filters

An NMR experiment can be viewed from either of two familiar points of view. According to the vector picture used throughout this book, an 90° rf pulse turns the equilibrium magnetization $\bar{M}_0$ into the x', y' plane. The motion of the observable component $M_{x,y}$ is then followed as the FID is detected. The frequency of rotation of the magnetization is the frequency of the line seen in the spectrum after Fourier transform of the FID. In the quantum mechanical picture (which was reviewed briefly in Section 2.4.2), the rf field causes transitions between energy levels and creates coherent motion of spins on the energy levels connected by the transitions. If the spin system is at equilibrium before the first pulse, the vector and the quantum mechanical pictures are complementary. The pulse creates coherences only between spins on those energy levels connected by a flip of one spin in the system (for allowed or single quantum transitions, $\Delta m = \pm 1$). The vector sum of the magnetic moments of all spins on the two levels is, because of their coherent motion, the observable magnetization component $M_{x,y}$. Since magnetization $M_{x,y}$ is a direct measure of the coherence between the two levels connected by a single quantum transition, it is often simply referred to as single quantum coherence or first-order (level) coherence.

An important distinction between the two pictures emerges if we study the effect of a pulse acting on a system that is not in an equilibrium state but already has some magnetization components in the x', y' plane. (Such nonequilibrium states are created by preceding pulses and delays.) We have followed many such situations in the earlier sections, and we have seen how magnetization components are rotated by the second pulse. In the quantum-mechanical picture the second pulse changes the first-order coherence in concert with the rotation of magnetization components in the vector description, but in addition, the second pulse creates coherences of higher orders. By a higher-order (level) coherence we mean the coherence between spins on levels separated by several quanta of energy. Such coherences occur if several spins flip simultaneously in a spin-spin coupled system. The order or level (p) of the coherence is the difference between the total magnetic quantum numbers of the spin system in the two energy levels involved (irrespective of the number of spins that have to flip). Thus a double quantum ($p = 2$) coherence connects states that differ in total magnetic quantum (m) number by 2, for example, a coherence between states $< +\frac{1}{2}, +\frac{1}{2}> $ ($m = +1$) and $< -\frac{1}{2}, -\frac{1}{2}>$ ($m = -1$) of an IS spin system is a double quantum coherence since $\Delta m = \pm 2$. In contrast, the coherence between states $< +\frac{1}{2}, -\frac{1}{2}>$ and $< -\frac{1}{2}, +\frac{1}{2}>$ would be a zero quantum coherence ($p = 0$) since $\Delta m = 0$, even though (as in the preceding case) two spins had to flip for the transition between the two states of the system. (Depending on whether the two flipping spins belong to the same sort of nuclei or not, the multiple quantum coherences are distinguished as homo- or heteronuclear; both play important roles in modern multiple pulse experiments.)

A higher-order coherence cannot be directly observed in pulsed FT NMR spectroscopy, but it can be converted into an observable first-order coherence and then studied indirectly. The conversion of higher-order (or multiple quantum) coherence back to observable single quantum coherence (magnetization) is also accomplished by an rf pulse.

All applications of multiple quantum coherences (or transitions) utilize some or all of the following properties, which depend on the coherence order p:

1. The sensitivity to magnetic field inhomogeneity is directly proportional to the coherence order p. In particular, zero quantum coherences are completely insensitive to magnetic field inhomogeneity.

2. Multiple quantum coherence evolution is not influenced by spin-spin coupling between the nuclei actively involved in the coherence (i.e., those which flip) but is affected by couplings with "passive" spins.

3. Multiple p-quantum coherence evolves under the influence of the algebraic sum of the chemical shifts of the involved nuclei. The sign of the chemical shift of the ith flipping nucleus in the sum is given by the direction of its flip (i.e., by Δm_i). In particular, double quantum coherence in a two-spin system evolves under the influence of the sum of the two chemical shifts, while zero quantum coherence in the same system evolves under the influence of the difference of the two chemical shifts.

4. Only an rf pulse can transfer coherence from one pair of energy levels to another pair of energy levels. The coherence transfer may change the coherence order by any integer number.

5. Multiple p-quantum coherence experiences p-fold the shift of phase of an rf pulse. In particular, if a rf pulse changes coherence of the order p_1 to coherence of the order p_2, it will follow that when the phase of the pulse is changed by ϕ, the phase of the p_2 quantum coherence changes by $(p_2 - p_1)\phi$.

6. The only directly measurable (detectable) coherences are single quantum coherences $p = \pm 1$.

7. At equilibrium, which is the state in which almost all experiments begin, the coherence order is $p = 0$.

Observations 2 through 7 form the basis of the method for systematic selection of coherence transfer pathways in NMR pulse experiments [186, 187]. By utilizing property 5, phase cycles can be designed so that the single quantum coherence eventually detected contains only those contributions that went through the prescribed coherence transfer pathway. All other contributions are added out. Using these concepts, the phase cycling required for quadrature detection in both dimensions and also for pure 2D absorption lineshapes (as well as other troublesome problems of 2D NMR) can be solved in a unified fashion. Since a detailed description of this general procedure is beyond the scope of this book, we shall use only observations 1 through 7 to explain the function of multiple quantum filters (which were developed [188–191] before the concept of the coherence transfer pathway was known).

Multiple quantum filters are pulse sequences that ensure by appropriate phase cycling that only those coherences that existed during a specified time period as p-quantum coherences are constructively added after detection. Thus a very useful homonuclear p-quantum filter can be added to the homonuclear COSY (Jeener) experiment. The filtration is achieved by adding to the Jeener sequence a third pulse in a short delay (to allow the spectrometer to reset the rf phase) after the second 90° pulse. The pulse sequence for this p-quantum filtered COSY then has the form

$$90^\circ_\phi - t_1 - 90^\circ_{\phi+\varepsilon} - 90^\circ_\varepsilon - \text{detection} \qquad (4.2\text{-}1)$$

where the phase ϕ is cycled through the values $\phi = k\pi/p$, $k = 0, 1, 2, \ldots$ $(2p - 1)$ for a p-quantum filter. The resulting signals are alternately added and subtracted (receiver phase cycling of $0, \pi$). For quadrature detection in the f_1 axis, phase cycling is also performed for $\varepsilon = 0$ and $\pi/2$ (as in COSY). The second pulse converts the magnetization that has evolved for evolution time t_1 not only into observable magnetization (that is detected in the Jeener experiment) but also into multiple quantum coherences. For example, if we choose to use the pulse sequence as a double quantum filter (DQF, $p = 2$), the phase ϕ is incremented by 90° in each consecutive pass through the sequence. The change of phase of the first two pulses by 90° produces a change of 180° in the phase of the double quantum coherence created by the second pulse (property 5 of multiple quantum coherence). The third pulse has only one purpose; it converts multiple quantum coherences into observable single quantum coherence. With the 180° phase change of the double quantum coherence, the observable single quantum coherence formed from the double quantum coherence also changes its phase by 180°. Hence the desirable signals detected in consecutive scans, which have to be added constructively, must be summed with alternating signs. Single quantum coherences also present after the second pulse change their phases by 90° as the phases of the first two pulses are changed. Then the observed coherence into which the single quantum coherence is converted by the last pulse also changes its phase by 90° in consecutive scans. Therefore, the undesirable signal detected in the third scan has an inverted phase, but by means of phase cycling of the receiver, this signal is added so that it balances the undesirable signal added in the first scan. Analogously, the undesirable signal of the fourth scan cancels the undesirable signal from the second scan. In this manner coherences that did not exist as double quantum coherences in the period between the second and the third pulse are eliminated.

The method of selection of the coherence transfer pathway can be depicted in a very lucid coherence transfer map (CT map). Such a map for the

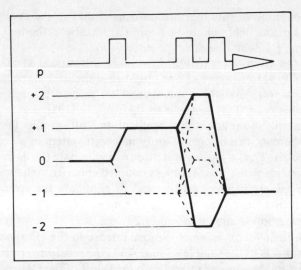

Figure 4.21 Coherence transfer map for a double quantum filtered H,H-COSY experiment, pulse sequence (4.2-1) with phase cycling given in the text. The experiment uses quadrature detection during t_2 as only coherence $p = -1$ is detected; the desired pathways (solid heavy lines) involve N-type peak detection since only coherence with $p = +1$ during time t_1 is allowed to reach detection; both pathways with $p = \pm 2$ during the mixing period are allowed; only coherences of the order $p \leqslant 2$ are shown; after the third pulse only the detected coherence is indicated).

case of a double quantum filtered COSY experiment is shown in Fig. 4.21. Several possible coherence transfer pathways are indicated by dashed lines, and the desired pathways are shown as solid lines. The phase cycling given for the pulse sequence ensures that only signals due to coherences that went through the desired pathway are constructively added in the computer memory.

The pathway starts with the coherence level $p = 0$ (property 7). Of the two coherences with $p = \pm 1$ created by the first pulse, we select only the one with coherence order $p = +1$ in order to ensure N-type peak detection (i.e., the direction of roation during the t_1 period is opposite to what will be detected during t_2). The coherence level will not change during the evolution time (property 4), although the coherence will evolve according to properties 2 and 3. The coherence is then converted by the second pulse into various coherences, and among them are coherences with $p = \pm 2$ which are converted after a short delay into the detected observable single quantum coherence of order $p = -1$ (quadrature detection with the opposite direction of rotation than during time t_1). According to property 1 the magnetization components that are defocused during the evolution by the effect of magnetic field inhomogeneities are being refocused during the detection time (N-peak or echo-type detection).

Multiple quantum filters (MQF) have been introduced as efficient techniques to simplify various types of one- and two-dimensional spectra. It should be noted, however, that the p-quantum filter is also transparent for odd multiple orders $p(2n + 1)$ where $n = 0, 1, 2, \ldots$ [191].

Most multiple quantum filter applications utilize the fact that a p-quantum coherence can be created only in a spin system of p coupled nuclei (spin-1/2 nuclei). Thus a p-quantum filter will eliminate the entire subspectra of all spin systems with $(p - 1)$ or less coupled spins from the spectrum; that is, a double quantum filter can be used to eliminate the strong signals of common solvents since they are usually singlets [191, 192]. An example of such an application is shown in Fig. 4.22.

Although the hydroxyl singlet is more intense in the 1D spectrum of this sample than it was in the sample of Fig. 4.13 (the mositure content is higher, as manifested by the accompanying high field shift of the OH resonance), the OH diagonal peak in the filtered 2D spectrum has only the intensity of the very weak cross-peak BC in Fig. 4.13; the extent of suppression is better demonstrated by comparison of the stacked-trace plots shown in Figs. 4.23 and 4.24.

The same principle is employed in the INADEQUATE method for suppression of strong single lines from isolated ^{13}C nuclei, but because of the importance of INADEQUATE, we shall discuss these experiments in a separate section to follow.

Similarly, by using a triple quantum filter we can eliminate from the spectrum all subspectra due to isolated two-spin systems, and so on. In general, with a properly chosen p-quantum filter, specific coupling patterns can be selected or suppressed. A weakly coupled spin system produces a diagonal peak in a p-quantum filtered spectrum if the spin in question has resolved couplings to at least $(p - 1)$ nuclei (which can be equivalent). A cross-peak in such a spectrum indicates not only that the two nuclei are coupled but that they couple with a common set of at least $(p - 2)$ additional spins.

Large values of p offer greater spectral simplification, but the sensitivity falls off rapidly with increasing filter order. Higher-order filters also require spectrometers equipped for small phase shifts of rf pulses.

The MQF can also be utilized in measurements of H,H-COSY (or NOESY) spectra with a pure absorption phase [193–195]. This combination is useful for determination of coupling constants and for identification of specific spin systems, especially in spectra that are otherwise crowded around the diagonal. Introduction of DQF results in spectra that have both cross-peak and diagonal multiplet peaks with an antiphase structure, and thus

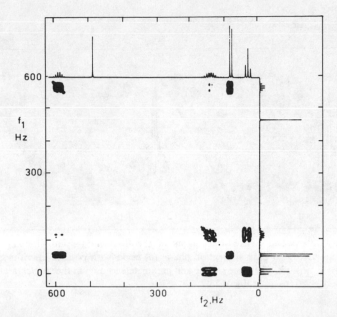

Figure 4.22 Double quantum filtered H,H-COSY spectrum of 2-butanol in deuteriochloroform; pulse sequence (4.2-1), experiment parameters as in Fig. 4.13 except that 32 transients were required for each t_1 increment for proper phase cycling, as described in the text; note higher H_2O content in this sample than in the sample shown in Fig. 4.13.

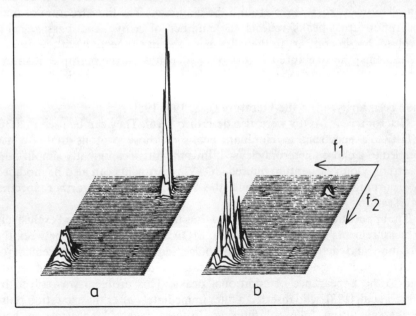

Figure 4.23 Stacked-trace plots of relevant parts of the spectra from (a) Fig. 4.13 (COSY) and (b) Fig. 4.22 (DQCOSY) (absolute-value presentation).

121

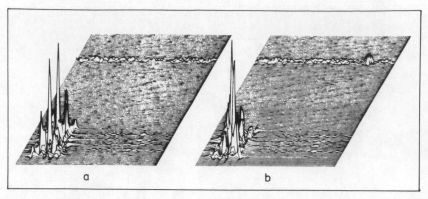

Figure 4.24 Whitewashed stacked-trace plots of the double quantum filtered H,H-COSY spectrum measured with pure absorption phase: (*a*) peaks with positive intensities; (*b*) peaks with negative intensities. The same sample and parameters as used in the spectrum of Fig. 4.22; same relevant positions as in Fig. 4.23.

cross-peaks near the diagonal can be diagnosed and analyzed. (We should recall that in COSY spectra the cross-peaks have absorption line shapes with the signs of peaks alternating within the multiplet, and the diagonal peaks have pure dispersion lineshapes. Together, these factors lead to considerable cancellation of cross-peaks around the diagonal.) The fine structure of pure absorption cross-peaks reflects the number of active and passive spins involved, but in contrast to the rules discussed in connection with unfiltered experiments, the structure of cross-peaks depends on the multiple quantum coherence transfer pathway. Several characteristic and pragmatically important examples of MQF and spin systems (including strongly coupled ones) have been analyzed in the literature [118, 194, 195].

Heteronuclear MQFs were also described [196]. They can be used in a 2D H,H-COSY spectrum to eliminate peaks of those protons that are not coupled to a chosen heteronucleus X; thus the filters can greatly simplify the spectrum. Double quantum filtered COSY experiments can also be modified to incorporate decoupling along the f_1 axis [197] for further spectral simplication.

The applications of MQF described thus far have been devoted exclusively to measurements of correlated spectra. MQF cannot be inserted between the evolution and detection periods of pulse sequences for measurement of resolved spectra, since the filter would cause undesirable mixing and would lead to the appearance of additional peaks. This problem was solved by Kessler et al. [198], who inserted a filter immediately after the excitation pulse so that evolution followed filtering. Filtered resolved spectra can help overcome problems in analyzing spectra of large molecules or spectra with strong singlet signals in a way analogous to the measurement of filtered correlated spectra.

4.2.2. INADEQUATE Spectra

We have mentioned on several occasions that some of the most difficult problems in structural determinations can be solved by use of the INADEQUATE method (Incredible Natural Abundance Double QUAntum Transfer Experiment). The INADEQUATE pulse sequence is essentially a double quantum filter applied to eliminate the strong singlet lines of isolated ^{13}C nuclei from ^{13}C NMR spectra and thus to make visible the weak satellite lines due to $^{13}C-^{13}C$ coupling.

The main advantages of conventional ^{13}C NMR spectroscopy over ^1H NMR are the large differences in chemical shifts of the different carbons present in the molecule and also the simple spectra produced. ^{13}C NMR spectra, which are most frequently measured with proton broadband decoupling, show only one line for each carbon, and the splitting due to homonuclear ($^{13}C-^{13}C$) coupling (analogous to the splitting due to $^1H-^1H$ coupling in ^1H NMR spectra) is missing in ^{13}C NMR spectra. Although this is a welcome feature in some respects (spectra are easier to interpret), we fail to obtain information about carbon-carbon connectivity which would otherwise be contained in the coupling constants. This spectral "simplification" occurs because of the low natural abundance (1.1%) of the carbon-13 isotope. To see the splitting due to $^{13}C-^{13}C$ coupling, the sample must contain molecules that have two adjacent carbon-13 atoms. Only 0.011% of all molecules in a sample without isotopic enrichment contain the adjacent pair of carbon-13 atoms that produces the ^{13}C satellites of a given line in the ^{13}C NMR spectrum. Hence coupling is manifested only as a pair of very weak satellite lines which are placed almost symmetrically around the strong central line which, of course, arises from molecules that contain only isolated single carbon-13 atoms (see Fig. 4.25).

The schematic shows the case in which the coupling constant is comparable to the chemical shift difference of the coupled nuclei; in other instances, with larger chemical shift differences, the intensities of the satellites are more evenly distributed.

To be able to measure these weak satellites in compounds that are not carbon-13 enriched, we must suppress the central line and improve the signal-to-noise ratio of the satellite lines by repeated accumulation of the FIDs until the lines are sufficiently distinct from the noise. Only then can we infer information about the coupling network from the satellites.

The INADEQUATE pulse sequence can be performed either as a one- or

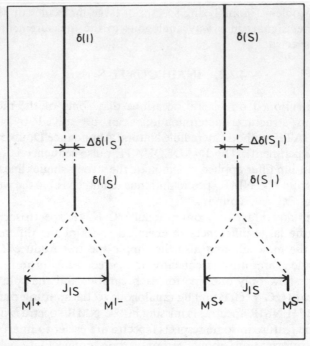

Figure 4.25 Schematic ^{13}C NMR spectrum of two neighboring carbons (I and S) measured with proton broadband decoupling: top, central lines from isolated carbons I and S; below, generation of the satellite spectrum by isotopic chemical shift effect and coupling between nuclei I and S. Intensities of the satellites are exaggerated, δ(I) and δ(S) are the chemical shifts of the two carbons, $\Delta\delta(I_S)$ and $\Delta\delta(S_I)$ are the isotope shifts from the presence of the neighboring ^{13}C nucleus, $^1J_{IS}$ is the direct coupling constant between the I and S nuclei; the satellites are labeled as in Fig. 2.19.

a two-dimensional experiment [188, 199–203]. The pulse sequence is given in Fig. 4.26 along with an oversimplified vector picture.

The delay d for weakly coupled spins is given [188] as

$$d = (2n + 1)/(4 \cdot J_{IS}) \tag{4.2-2}$$

in order to have optimum coherence transfer. Usually, the value of n is chosen to be zero in order to avoid excessive relaxation during the delay. Other values of n can be used in order to have a delay d which would simultaneously be optimal for different coupling constants J_{IS} (e.g., to observe simultaneously double quantum coherences due to one- and two-bond ^{13}C–^{13}C couplings).

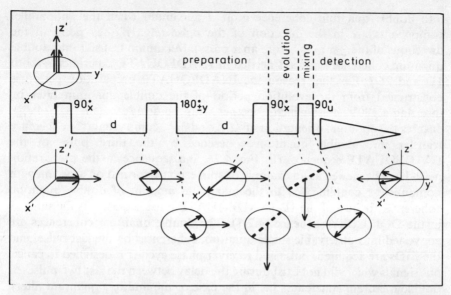

Figure 4.26 2D INADEQUATE pulse sequence and evolution of the magnetizations corresponding to one central line and its two satellite lines. Protons are continuously broadband decoupled, magnetizations of M^{1+} and M^{1-} satellites are exaggerated, frequency of the rotating frame is set equal to the frequency of the central line, only the evolution for $t_1 = 0$ is shown, dashed heavy line schematically denotes double quantum magnetization; optimum $d = 1/(4 \cdot J_{IS})$; for 1D INADEQUATE experiments the evolution time t_1 is constant for a short delay time, e.g., 10 ms.

Although INADEQUATE is older than the more general multiple quantum filters discussed in the preceding section, the INADEQUATE method is easy to understand as an example of a double quantum filter especially suited or optimized for $^{13}C-^{13}C$ double quantum coherence. Also, because of the low natural abundance of ^{13}C, no order higher than double quantum coherence can interfere. Each $^{13}C-^{13}C$ pair is isolated from any other ^{13}C nucleus, and the pair can be treated as an isolated IS spin system.

Let us consider first the one-dimensional INADEQUATE experiment, which uses the pulse sequence from Fig. 4.26 (the only difference is that time t_1 is kept constant and short). Except for the refocusing 180° pulse in the midpoint of the preparation period and different timing intervals, the INADEQUATE pulse sequence is the same as the sequence (4.2-1) for the double quantum filtered COSY, but the functions of the various periods are different. In two-dimensional COSY the amount of coherence that is transferred into double quantum coherence by the second pulse varies periodically with evolution time t_1, and it is at a maximum when the multiplet components are antiparallel before the second pulse. The transfer into double quantum coherence is at a maximum when the antiparallel components are in the direction of the pulse (e.g., if they point in the directions of the $\pm x$ axes before an x pulse). Maximum transfer into double quantum coherence is always needed in INADEQUATE experiments (both 1D and 2D). For that reason the INADEQUATE preparation period is constructed from the evolution period of the double quantum filter by inserting a $180^\circ_{\pm y}$ pulse which, together with the delay set as $d = 1/(4 \cdot J_{IS})$, ensures the optimum orientation of M^{1+} and M^{1-} components for coherence transfer into double quantum coherence by the third pulse of the INADEQUATE sequence. (In Fig. 4.26, the sequence in the preparation period can be viewed as an example of spin echo performed on a system with homonuclear coupling.) With the maximum amount of double quantum coherence, the rest of the 1D INADEQUATE pulse sequence is the same as in the DQF COSY sequence (4.2-1). The double quantum coherences are converted into observable single quantum coherences by the last pulse, and the FIDs are acquired. Pulse and receiver phase cycling is designed to cancel out signals which did not exist during the delay between the last two pulses as double quantum coherences (as in the case of the multiple quantum filter). The two magnetization components M^{1+} and M^{1-} appear in the antiparallel arrangement after the last pulse; thus the satellite lines have opposite phases in the 1D INADEQUATE spectrum. An example of the 1D INADEQUATE spectrum of 2-butanol is shown in Appendix A.9 in Fig. 6.17.

Exact values of $^{13}C-^{13}C$ coupling constants can be determined from 1D INADEQUATE spectra, and the coupling network can be established by picking out the repeated spacings. The task of comparing the various spacings in the spectrum can be performed by a computer [204], but even with the help of a computer, the coupling network cannot be determined unambiguously if the magnitudes of coupling constants are not sufficiently well differentiated or if the molecule is too complex. In such a case, the 2D version of INADEQUATE is of considerable help. 2D INADEQUATE identifies the two coupled nuclei, not by the value of their common coupling, but by the frequency of their common double quantum coherence.

The pulse sequence for two-dimensional INADEQUATE is easily constructed from the 1D sequence by incrementing the delay between the last two pulses in the same way that the evolution time t_1 is incremented (in all other 2D experiments). The maximum possible amount of double quantum coherence created at the end of the preparation period evolves during time t_1. The evolution proceeds according to observations 2 and 3 from the preceding section. The evolution is not affected by J_{IS} coupling between the two carbon-13 nuclei I and S, and it rotates with a frequency given by the sum of their chemical shifts. The double quantum coherence is turned into single

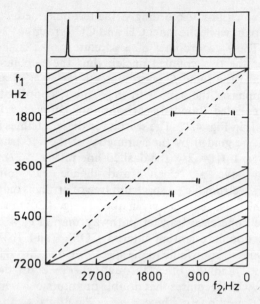

Figure 4.27 2D INADEQUATE (contour plot) and 1D ^{13}C NMR spectra of 2-butanol in hexadeuteriobenzene; pulse sequence of Fig. 4.26 modified as described in the text for N-type peak detection, relaxation delay between two scans 12 s, spectral width in f_2 axis 3600 Hz, in f_1 axis 7200 Hz, 128 transients accumulated for each of 96 increments of t_1, delay $d = 6.7$ ms, continuous broadband decoupling of protons, FID data matrix 96 × 1024, 2D spectrum matrix 512 × 2048, power spectrum presentation; the diagonal is indicated by the dashed line, the horizontal solid lines connect the peaks of coupled nuclei; for explanation of the shaded area, see the text.

quantum coherences by the final pulse, and the effectiveness of this last coherence transfer depends on the orientation of the double quantum coherence with respect to the magnetic field of the final pulse. Of course, the orientation varies periodically in time t_1 with the rotating frequency of the double quantum coherence. The spectra obtained after the first Fourier transform for each of the t_1 values show the satellite lines as in 1D INADEQUATE spectra (with the central lines suppressed), and the spectra vary with t_1. After the second transform, the satellite peaks have f_1 frequencies equal to the sum of the chemical shifts of the coupled (neighboring) ^{13}C nuclei [$\delta(I) + \delta(S)$]. An example of the 2D INADEQUATE spectrum of 2-butanol is shown in Fig. 4.27.

Obviously, interpretation of this spectrum in terms of carbon-carbon connectivities is straightforward. One or more satellite pairs belong to each line of the usual 1D NMR spectrum. The number of the satellite pairs belonging to a given 1D line indicates the number of carbon atoms directly bonded to the given carbon. Accordingly, the terminal methyl carbons have only one satellite pair, while the inner CH and CH_2 groups of 2-butanol have two pairs each. The f_1 coordinate of a satellite peaks is the sum of the chemical shifts of the two coupled nuclei; thus the satellite peaks of the coupled nuclei have the same f_1 coordinates $[\delta(I) + \delta(S)]$ and can be so identified. For emphasis, the pairs with the same f_1 coordinate, that is, the satellites of the coupled (bonded) carbons, are connected by solid lines parallel with f_2 axis in Fig. 4.27. The search for these satellites in crowded or noisy spectra can be guided by the symmetry of the spectrum around the "skew diagonal" with $f_1 = 2f_2$ (the dashed line in Fig. 4.27).

In order to have a "clean" and hence easy-to-interpret 2D INADEQUATE spectrum (i.e., a spectrum free of artifacts and image peaks, as in Fig. 4.27), the pulse sequence given above (Fig. 4.26) must be modified. According to the coherence transfer pathway map (Fig. 4.21), the phase cycling used for double quantum filtered COSY (and INADEQUATE) allows two pathways to contribute to the spectrum, one through coherence of the order $p = +2$ and the other through $p = -2$. In the case of 2D INADEQUATE, which requires that double quantum coherences evolve for evolution time t_1, only one of these $p = \pm 2$ pathways can be allowed if quadrature detection is to be used along the f_1 axis (unless special data handling is used to obtain pure absorption line shapes).

We have already seen how important advantages are gained by using quadrature detection in measurements of 2D NMR spectra. Using single-channel detection in homonuclear correlations, the observing transmitter should be set at the end of the spectrum. The resulting loss in experiment sensitivity and efficiency (because of inefficient use of transmitter power, memory, and data handling) is amplified in double quantum spectroscopy since the spectral range along the f_1 axis is doubled relative to that along the f_2 axis. More memory space is wasted, and more t_1 increments are needed for the same resolution, thus degrading the overall performance of the experiment if quadrature detection cannot be employed. If the INADEQUATE spectrum, is measured without quadrature detection along the f_1 axis, although quadrature detection is employed in the detection period with the transmitter (and receiver) frequency positioned into the middle of the spectrum, the resulting 2D spectrum contains various images and reflections of the folded peaks, and the interpretation of the spectrum of a molecule of even moderate complexity could pose a problem.

As in other correlations, quadrature detection along the f_1 axis requires elimination of one of the two counterrotating components in the evolution period. In the case of INADEQUATE the two counterrotating components are the two double quantum coherences with the order $p = +2$ and $p = -2$. To detect N-type (echo) peaks the modification of the pulse sequence must ensure that only double quantum coherence of the order $p = +2$ reaches the receiver (after conversion into single quantum coherence) with the appropriate phase to be added constructively to the memory. Two quadrature detection schemes have been proposed by Freeman et al. [201, 202]. The first variant uses a more complicated pulse sequence that rotates the single quantum magnetizations through 45° around the z axis, while the double quantum coherences are rotated through 90°. Proper phase cycling then ensures cancellation of unwanted signal components [201]. The second variant is easier to implement. It uses the INADEQUATE pulse sequence of Fig. 4.26 except that the final "read" pulse is allowed to have an arbitrary flip angle α,

$$90^\circ_x - d - 180^\circ_x - d - 90^\circ_x - t_1 - \alpha_x - \text{detection} \qquad (4.2\text{-}3)$$

with the optimum setting of delay d noted above and broadband proton decoupling throughout.

The discrimination between the two double quantum coherences is achieved on the basis of the signal dependence on the flip angle α. When a 90° flip angle is used, the two double quantum coherences contribute equally to the observed signal, the direction of rotation cannot be determined, and the observed signal is amplitude modulated. If the "read" pulse flip angle is greater or smaller than 90°, the two double quantum coherences contribute to the time-averaged spectrum to a different extent. Good discrimination between the two components is obtained with $\alpha = 60°$ (anti echo) or $\alpha = 120°$ (echo), the latter being preferred, since in the case of echo or N-type peak detection the effects of magnetic field inhomogeneity are partially refocused. An additional and welcome feature is an approximately 30% increase in signal intensity when the flip angle has one of the two foregoing values (as compared to a signal obtained with a 90° pulse). This modification of the INADEQUATE pulse sequence was used for measurement of the spectrum shown in Fig. 4.27.

When 2D INADEQUATE spectra are measured with quadrature detection, they have the very useful symmetry described above. Since the INADEQUATE peaks have the coordinates $f_1 + f_S$, f_1 or f_S, they cannot be found in the parts of the spectrum shown as shaded areas in Fig. 4.27. Thus one half of the data matrix, i.e., the memory locations that contain information about a signal with frequencies in the shaded area, is wasted. The memory can be put to better use if the spectrum is allowed to be folded in the f_1 direction. This folding is accomplished by incrementing the t_1 time by 1/SW (where SW is the spectral width of the 1D spectrum instead of by 1/(2·SW). The area around the skew diagonal is then folded back as shown in Fig. 4.28.

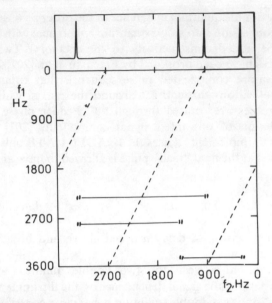

Figure 4.28 2D INADEQUATE spectrum of 2-butanol in hexadeuteriobenzene; pulse sequence 4.2-3 with N-type peak detection, $\alpha = 120°$, experimental parameters as in Fig. 4.27, except for the spectral width in the f_1 dimension which was 3600 Hz, the dashed line is the diagonal that is broken by folding along the f_1 axis.

With the t_1 sampling rate reduced by half, the number of increments may also be halved while retaining the same resolution in the f_1 direction. The resulting saving of time allows the number of scans for time averaging of each increment in t_1 to be doubled to give the same sensitivity. The savings in data storage requirements allows the digital resolution to be doubled in the f_2 direction at negligible costs in time but with additional possible improvement in signal-to-noise ratio [205–208]. These folded spectra, however, make it more difficult to apply mathematical noise reduction by spectrum symmetrization around the skew diagonal (analogous to the symmetrization of COSY spectra described earlier [209]).

Despite these important improvements, sensitivity remains the main problem of INADEQUATE experiments. Therefore, it is not surprising that variants of INADEQUATE that employ INEPT [210] or DEPT [211] for signal enhancement by polarization transfer from protons have been proposed. Unfortunately, most important applications of INADEQUATE are concerned with quaternary carbons, and in those cases INEPT or DEPT can only utilize small couplings over two or three bonds. A considerable increase in sensitivity should be gained by the INSIPID method (INadequate

Sensitivity Improvement by Proton Indirect Detection) [212], which requires a reversed probe to allow detection of the ^{1}H signal while ^{13}C resonances are irradiated by rf pulses.

Another important modification is the uniform excitation of double quantum coherences, even in the case of nonuniform coupling constants, which permits differentiation between directly coupled and remote pairs of carbons [213]. Spectral editing according to carbon line multiplicity has also been described [211, 214] but has found little use.

Naturally, analogous double (or multiple) quantum correlated 2D spectra can be measured for other coupled pairs of nuclei (e.g., ^{1}H–^{1}H). Such measurements greatly facilitate analysis of spin systems in crowded spectra of complex molecules such as proteins [215–218].

In this connection a useful and interesting modification called DOUBTFUL was proposed by Hore et al. [190] to obtain the spectrum of selected multiplets within a crowded region of the spectrum. DOUBTFUL produces a simpled 1D spectrum by averaging ^{1}H,^{1}H-INADEQUATE spectra with different t_1 times. The coaddition leaves only the spectra of the multiplets selected by the setting of transmitter frequency [190, 219, 220].

The success of these experiments depends to a considerable extent on the rf pulses being free of errors (the most common errors are pulse inhomogeneity over the sample and off-resonance effects). The errors can be minimized by the use of compensate pulses, but correct composite pulses should be employed in a proper way [221–223]. Incorrectly compensated pulses could even produce a worsening of the experimental results [224].

In summary, 2D INADEQUATE is a very powerful method for structure determination, but because of low sensitivity the experiments must be carefully prepared and probably optimized (maximum digital resolution in the f_2 axis and minimum number of t_1 increments with an acceptably degraded resolution in the f_1 direction). The experiments require high sample concentration (not always compatible with resolution and relaxation), which might require an increase of the sample temperature. The length of the relaxation period before the first pulse of the sequence is also important. It can be shortened by addition of a relaxation reagent [such as chromium(III) or iron(III) acetylacetonate], but the relaxation time should not be shorter than the duration of the pulse sequence. In any case the presence of appropriate peaks in the INADEQUATE spectrum can be taken as positive evidence for coupling between two nuclei, but failure to see some expected peak should not be taken as an unequivocal denial of the proposed structure since it is possible that the experimental conditions were not properly set for observing this particular double quantum coherence (e.g., as in the case of strongly coupled carbons [225] or in the case of anomalous values of coupling constants).

4.3. 2D NMR EXCHANGE SPECTRA

The correlation experiments discussed so far have all been based on magnetization transfer between nuclei that were spin-spin coupled. The magnetization transfer took place with the same phase and orientation throughout the sample; it was a coherent transfer. The resulting spectrum shows which nuclei are spin-spin coupled and thus provides information on molecular structure.

Magnetization transfer can also be accomplished in an incoherent fashion. Incoherent magnetization transfer accompanies dynamic processes such as chemical exchange and dipole-dipole relaxation. These processes cause magnetization transfer between different species as the nuclei are exchanged or as they interact. The exchange (or interaction) takes place at various moments and at various locations in the sample; hence the magnetization transfer is incoherent. The spectra identify the species or sites between which exchange or dipole-dipole interaction occurs.

The study of dynamic processes by nuclear magnetic resonance has long been known to provide a unique insight into molecular dynamics because it permits the observation of a system in an equilibrium condition, whereas other methods such as stopped flow or temperature jump place the system (at least initially) into a chemically nonequilibrium state [226].

Various dynamic processes that are studied by NMR are rather loosely termed "chemical exchange" in the literature, and the branch of NMR spectroscopy concerned with the study of such processes is called "dynamic NMR" or "exchange NMR" spectroscopy. We shall follow this example, and the heading "2D NMR Exchange Spectra" is used in this broad sense.

In addition to true chemical exchange, in which an exchange of atoms or groups between different molecules takes place (as in the literal exchange of hydroxyl protons in solutions of alcohols), chemical exchange also includes such physical processes as internal rotation in which no chemical bond is formed or broken. We shall also include measurements of the nuclear Overhauser effect (NOE) in this section on 2D exchange spectroscopy. Since NOE also involves incoherent magnetization transfer via dipole-dipole interaction, NOE is measured by a method similar to that used for true chemical exchange, but provides information about spatial proximity of various parts of the molecules.

In addition to the many facets of 2D NMR spectroscopy that we have already discussed, we must also consider exchange rates (or more exactly, relative exchange rates). The relative exchange rate is the rate of the exchange process measured by the frequency difference, Δf, of the lines that belong to the molecular species between which the exchange takes place. (For example, Δf for the internal rotation around the C—N bond in N,N-

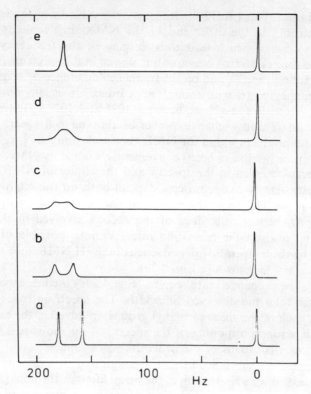

Figure 4.29 Temperature dependence of the ^{1}H NMR spectrum of N,N-dimethylacetamide; 50% solution in hexadeuteriobenzene; the frequency of the CH_3C line is taken as an arbitrary origin of the frequency scale; temperatures: a, 23°C; b, 60°C; c, 70°C; d, 75°C; e, 90°C; Δf = 21.6 Hz.

dimethylacetamide is found to be 21.6 Hz when it is followed by ^{1}H NMR spectroscopy of the methyl groups at 200 MHz (see Fig. 4.29). The relative rate of the exchange process determines which of the spectral methods is suitable to study the process.

Let us recall briefly some fundamental points. According to the Heisenberg uncertainty principle, the finite lifetime of a molecular species causes broadening of the spectral line. The broadening contribution Γ (in hertz) is given as

$$\Gamma = 1/(\pi \cdot \tau) \qquad (4.3\text{-}1)$$

where τ is the mean lifetime of the species (in seconds). Since the usual linewidth of a high-resolution NMR line is in the range 1 to 10^{-1} Hz, a lifetime τ shorter than 1 to 10^{-1} s leads to observable line broadening in the NMR spectrum. At the other end of the NMR time scale are very fast exchange processes, which lead to averaging of the frequency difference between the species, so that only one line is seen in the spectrum. Depending on the relaxation time T_2^* and on the frequency difference Δf being averaged by the exchange, spectra with a simple appearance are usually observed when the mean lifetime is somewhere between 10^{-4} and 10^{-6} s. There are many chemically interesting exchange processes that have a mean lifetime of reactants and products within the NMR time scale limits of 1 to 10^{-6} s; they therefore can be (and have been) conveniently studied by NMR techniques.

The effects observed in the spectra and the choice of NMR method to study a particular exchange process depend both on the relative exchange rate (i.e., on the lifetime of the molecular species, τ, and on the frequency difference, Δf, between the lines of the species involved in the exchange process) and also on the relaxation rates. Various possible situations are illustrated by the temperature dependence of the ^{1}H NMR spectrum of N,N-dimethylacetamide (shown in Fig. 4.29).

In the "slow exchange" rate regions (Fig. 4.29a) the line separation Δf is much larger than the observed linewidth. The spectrum exhibits separate signals for each of the species (methyl groups) involved in the exchange; the spectrum is a superpositioning of the spectra of the individual species. The mean lifetime τ is usually of the order 10^{-2} s or longer.

As the lifetime is shortened further, the region of "intermediate exchange" rate is entered. With a shortening of the mean lifetime, the lines grow broader and broader, and the apparent maxima of the lines move toward one another until coalescence occurs (Fig. 4.29b). Usually, an intermediate exchange includes lifetimes in the region 10^{-2} to 10^{-5} s.

After coalescence, a further increase of the exchange rate leads to line narrowing. In this "fast exchange" rate region, NMR spectroscopy is measuring too slowly and "sees" only one averaged species.

Different techniques must be used for different rates of exchange processes. Very fast systems can be studied through their effect on nuclear relaxation. Intermediate processes can be studied by lineshape analysis. Slow exchange processes can be characterized by monitoring the transfer of nuclear polarization between nonequivalent sites and are particularly amenable to the 2D exchange techniques to be discussed here.

The main advantage of 2D NMR studies of exchange processes is that all exchange sites can be followed simultaneously, and the high resolution that is available in 2D NMR spectroscopy aids in the resolution of these sites. These advantages have been amply demonstrated in the work of Wüthrich et al., who have utilized the NOESY technique (elaborated in the second part of this section) to elucidate the structure of proteins [227–229].

We shall focus first on the basic homonuclear 2D exchange experiment and then later consider some of its most important modifications.

4.3.1. 2D NMR Observation of Exchange Processes

A general method for 2D NMR spectral studies of exchange processes has been demonstrated by Jeener et al. [230]. All pulse sequences of exchange spectroscopy are, in essence, derived from the basic homonuclear pulse sequence [230, 231] depicted in Fig. 2.4c and repeated here with the relative phases of nonselective pulses specified to allow discussion of the mechanism of the experiment:

$$d_{prep} - 90_x^\circ - t_1 - 90_x^\circ - \text{mag grad}, d_m - 90_x^\circ - \text{detection} \qquad (4.3\text{-}2)$$

Here, d_{prep} is the relaxation delay shown explicity, and the mixing delay d_m contains a pulse of magnetic field gradient.

We shall follow the action of this pulse sequence on the "exchange" of two N-methyl groups by hindered rotation around the N—C(O) bond in N,N-dimethylacetamide:

$$\begin{array}{ccc}
\text{Me} & & \text{Me} \\
\diagdown & & \diagup \\
& \text{N—C} & \\
\diagup & & \diagdown\diagdown \\
\text{Me} & & \text{O}
\end{array}$$

Transverse magnetizations are created at the end of the preparation period by the first 90° pulse of the sequence (4.3-2). The three magnetization vectors, each of which belongs to the protons of one of the three nonequivalent methyl groups, rotate with different frequencies in the x', y' plane. The rotational frequencies correspond to the chemical shifts of the nonequivalent protons. During the evolution period the magnetiztions travel different angles, or, in 2D terminology, they become frequency labeled. The second 90° pulse, which is also applied along the x' axis, selects the y' components of the precessing magnetizations and rotates them into the direction of the z axis. The remaining transverse magnetization components could interefere with the final observed FID, and therefore they must be eliminated. In the basic pulse sequence the elimination is achieved by defocusing the undesired magnetization components by the magnetic field gradient pulse.

The z-axis components form the initial state for the mixing or exchange process. Their magnitudes carry the frequency labels from the evolution period; the magnitudes are determined by the precession angles that the magnetizations acquired during the evolution period t_1, and thus they are marked during t_1 as the components belonging to a particular type of nonequivalent protons. The exchange process to be investigated by the experiment occurs during the mixing period. The mixing period is of fixed length, d_m; all through its duration the magnetizations remain longitudinal along the z axis.

The two nonequivalent amidic methyl groups are exchanged by the hindered rotation around the N—C(O) bond. The exchanging methyl groups take with them their respective "labeled" z-axis magnetization components. Magnetization transfer by this incoherent mechanism requires a certain period of time to take place; thus, to have measurable exchange effects, the mixing time d_m must be comparable to the reciprocal of the rate of the exchange process studied.

After the exchange, the third 90° pulse at the end of the mixing period rotates the longitudinal magnetizations (z-axis components) into the x', y' plane for detection and recording during t_2. The magnetizations rotate during the detection period with the frequencies corresponding to the chemical shifts of the three nonequivalent protons. Their magnitudes contain contributions from magnetizations that were exchanged during the mixing period and which were labeled by different frequencies during the evolution period. As in all 2D NMR measurements, the experiment is repeated for a number of equally spaced evolution times t_1. After the first Fourier transform we obtain spectra with intensities modulated in amplitude to an extent proportional to the amount of magnetization transferred; the frequency of modulation of a line is the frequency of the magnetization from which the magnetization was transferred during the mixing period. The frequency is apparent from the 2D spectrum obtained after the second Fourier transform. An example of the 2D exchange spectrum (in contour form) of N,N-dimethylacetamide is shown in Fig. 4.30. Typical experimental parameters are provided.

The diagonal peaks in a 2D exchange spectrum identify the individual protons in the molecule as in a 1D spectrum, since the diagonal peaks are produced by those protons which have not been exchanged during the mixing time. The cross-peaks connect the diagonal peaks of the products with those of the reactants and indicate that exchange took place between the corresponding sites. That is, the appearance of an off-diagonal peak at frequencies (f_1, f_2) in the spectrum indicates that an exchange process during the mixing period has transferred magnetization components of precession frequency f_1 to a transition of frequency f_2. In a very general sense, the resulting 2D spectrum can be regarded as showing the spectrum of the reactants along the f_1 axis and the spectrum of the products along the f_2 axis. The cross-peaks indicate which reactants lead to which products.

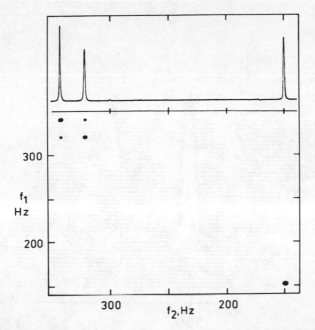

Figure 4.30 2D exchange and 1D ¹H NMR spectra of *N,N*-dimethylacetamide in hexadeu-teriobenzene (50%); pulse sequence of Fig. 2.4*c*, sample temperature 30°C, mixing time 0.5 s, preparation period 5 s, 8 transients for each of 512 t_1 time increments, FID data matrix 512 × 1024, 2D spectrum matrix 1024 × 1024, spectral width 500 Hz in both dimensions, power spectrum presentation.

In Fig. 4.30 the positions of the cross-peaks demonstrate that only exchange between amide methyl groups takes place. No exchange with the acetate methyl protons is detected. (A cross-peak between acetate and amide methyl lines would indicate a nuclear Overhauser effect between the groups.) A more quantitative picture about the rate of exchange can be obtained from the stacked trace plot of the detailed spectrum shown in Fig. 4.31.

The technique described is a general one, and it can be used to study true chemical exchange as well as the exchange of magnetization by dipole-dipole interaction. The pulse sequence is the same as that used for making correlations by means of the dynamic nuclear Overhauser effect known as NOESY [232, 233] (by analogy with COSY).

With increasing exchange rates, 2D exchange spectra behave the same as 1D NMR spectra. That is, as the relative rate of exchange becomes higher, the peaks broaden, move closer together, and eventually coalesce into a broad peak that subsequently narrows as shown for 1D NMR spectra in Fig. 4.29.

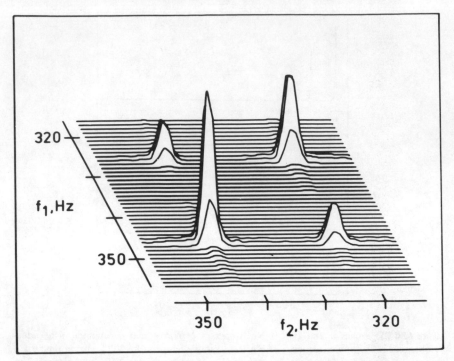

Figure 4.31 Whitewashed stacked-trace plot of a part of the 2D exchange spectrum of *N,N*-dimethylacetamide. For experimental details see Fig. 4.30; note that the figure has been rotated by 90° so that the diagonal is going from the lower left to the upper right corner.

The largest amount of information can be gained from 2D NMR measurement when it is applied to systems that fit the category of slow exchange. In such cases the cross-peaks not only indicate exchange but the intensities offer information on the rate of exchange and on the exchange pathways as well. The 2D exchange technique is of particular value not only for slow exchange reactions between two species or sites but also for the detection of exchange networks in systems with many potential sites of exchange, because it is possible in a single 2D NMR experiment to label all sites and thus all exchange pathways can be simultaneously observed [234]. This is particularly useful, of course, in determining the structure of macromolecules such as those found in biopolymers or molecular biology systems.

It should be noted that the intensities of cross-peaks are not necessarily simply proportional to the elements of the kinetic matrix. Several precautions must be taken to obtain reliable quantitative data on exchange rates or on nuclear Overhauser effects from 2D exchange spectra. Some of these precautions are briefly reviewed in the next section.

4.3.2. Quantitative Evaluation, Modified Pulse Sequences, NOESY

The description of the basic 2D exchange experiment in the preceding section reflected the overly optimistic view of these experiments that was generally shared at the outset of 2D exchange spectroscopy. Later analyses of the experiments have shown that to obtain quantitative data, the experiment must be planned very carefully; the basic pulse sequence (4.3-2) must be modified, acquistion parameters (d_m) carefully chosen, and the results given proper mathematical treatment. We shall mention briefly some of the considerations involved, as well as some solutions, but it is not possible here to discuss them all in detail.

Even in early experiments [230] it was obvious that the use of a magnetic field gradient pulse to eliminate transverse magnetizations is not the best solution because it introduces a disturbance of the field frequency/look system. It was suggested [230] that phase cycling of the rf pulse be used instead, with the same final effect.

To design the phase cycling along the lines of the preceding section would be simple. It would suffice to compensate in alternative scans the magnetization components remaining in the x', y' plane after the second pulse. However, only seldom do the studied spin systems consist solely of isolated nuclei; they often contain coupled spin systems.

In spin-spin coupled systems, the first two pulses of sequence (4.3-2) creates coherences of various orders (as discussed in Section 4.2.1) which are turned into observable magnetizations by the third pulse. As a result, 2D exchange spectra are complicated by the presence of what are termed "J cross-peaks." The J cross-peaks produced by this coherent magnetization transfer are undesirable in exchange spectra since they may be mistaken for genuine exchange cross-peaks or they may render the measured intensities inacurate in the case of coincidence with true exchange cross-peaks.

The J cross-peaks can be identified by comparison with COSY spectra (which contain J cross-peaks exclusively) or by comparison of their phase properties in 2D exchange spectra. Not only do the J cross-peaks have vanishing net intensity (i.e., the peaks in the multiplets appear with alternating intensity signs), but their phase is also different from the phase of the exchange peaks. (When the phases are adjusted to show absorption exchange peaks, the J cross-peaks become dispersive [235].) Naturally, it is necessary to present the spectrum in a phase-sensitive mode in order to see these differences. Especially when investigating large molecules, it is advantageous to have measurement schemes that produce pure absorption phase spectra (Sections 2.2 and 6.7) [102, 187, 236, 237].

Phase cycling schemes can be designed to eliminate not only the magnetizations (i.e., the first-order coherence) remaining in the x', y' plane after the second rf pulse but all higher-order coherences as well [235, 238, 239]. (Simultaneous elimination of axial peaks is an additional advantage of phase cycling.) Elimination of the zero-order coherences can be achieved by digital filtering [235], by a modification of the pulse sequence (essentially an insertion of a 180° pulse, by the varying of its position within the mixing delay d_m) [239, 240], or by varying the mixing time d_m[234, 235, 239, 241, 242].

Other possible sources of inaccuracy are "t_1 noise" and "t_1 ridges," which are due to instrumental instabilities and digitalization errors or baseline distortions. They can also be eliminated by phase cycling [36].

We have already mentioned that successful experiments require a proper choice of the mixing time. The optimum mixing time d_m which leads to maximum intensity of the exchange cross-peak can be estimated according to the formulas available for some simple systems [230, 238, 243]. It is necessary to consider not only the appropriate exchange rate but also the longitudinal relaxation time T_1 since the longitudinal magnetization components that are being exchanged also relax at the same time. Only in the simplest systems (such as two-site exchange (A $\rightleftharpoons$ B), with equal populations, equal relaxation rates, and described by the single exchange rate constant k) can the constant (which is independent of the relaxation rate [243]) be determined directly from the ratio of cross-peak (I_{AB}) and diagonal peaks (I_{AA}) intensities:

$$\frac{I_{AA}}{I_{AB}} = \frac{1 + \exp(-2kd_m)}{1 - \exp(-2kd_m)} \qquad (4.3\text{-}3)$$

For short mixing times the formula can be approximated by the simple ratio $(1 - kd_m)/kd_m$, which shows that maximum exchange effects are observed when the mixing time is close to the reciprocal of the rate constant. In many situations, however, Eq. (4.3-3) is not practical. It is often difficult to measure the amplitudes of the diagonal peaks because the resonances may be hidden in a not-well-resolved "ridge" along the main diagonal. Also, it often happens that several exchange mechanisms combine to form extensive exchange networks. In such cases it is not possible to derive exchange rates from simple ratios, and it becomes necessary to record a series of 2D exchange spectra with different d_m values, including some very short ones [242]. A numerical evaluation of the d_m dependence of the cross-peak intensities yields a quantitative measure of the rate constants involved. Recording the required set of 2D exchange spectra for various times d_m represents, in effect, an extension from 2D to 3D spectroscopy. Although conceptually simple, it is exceedingly demanding on experimental time and data storage. These demands can be very significantly reduced by employing the "accordion spectroscopy" introduced by Bodenhausen and Ernst [243]. In accordion spectroscopy the mixing time is varied simultaneously with the evolution time:

$$d_m = \kappa t_1 \tag{4.3-4}$$

The proportionality reduces the 3D experiment to a special form of a 2D experiment, and the second Fourier transform with respect to t_1 is at the same time a transform with respect to d_m. In the resulting spectrum, the f_1 and f_m axes run in parallel, but the spectral widths differ by a factor of κ. The quantitative analysis is simple since the linewidths of the resulting Lorentzian lineshapes provide a direct measurement of the rate constants of the particular exchange process (whether it is chemical exchange, NOE, or spin-lattice relaxation) [243]. When accordion spectroscopy is applied to the study of chemical exchange using ^{13}C NMR spectroscopy, J cross-peaks are no problem. The method is less well suited, however, to ^1H NMR because of zero quantum effects [244]. (The idea of synchronously changing some other parameter with t_1 has now also been extended to some other experiments; for example, the flip angle has been varied in an H,X-COSY [110] experiment.)

Nuclear Overhauser Enhancement SpectroscopY (NOESY)

The nuclear Overhauser effect (NOE) is a useful source of information for chemical structure elucidation [53] because it is one of the few quantities that yields information on interatomic distances in solutions. The NOE is measured as the fractional change in intensity by cross-relaxation of one NMR line when another resonance is perturbed by irradiation. Because of the inverse sixth-power dependence of the dipolar interaction on the internuclear distance, the steady-state NOE is a sensitive measure of the distance between observed and perturbed nuclei and has frequently been used in structural studies of small molecules. For determination of relative internuclear distances, the small positive NOEs must be measured with a high relative accuracy that requires low repetition time (relaxation delays of three to five times T_1 are necessary). For these reasons, 2D measurements of NOEs in small molecules are not very practical. The power of a 2D measurement of NOE lies in its application to large molecules.

For macromolecules the practical use of 1D NOE measurements is limited because of the need for long accumulation times and the poor selectivity for irradiation of individual lines in crowded portions of a 1D spectrum. The problem of selectivity does not exist when a 2D technique is employed. All resonances are perturbed by the nonselective 90° pulse, but the perturbations are frequency labeled through time t_1. Thus NOESY can provide a complete set of NOEs in the molecule, with a single instrument setting.

It is obvious that 2D NOESY [245] is a very efficient method. Moreover, since a negative NOE and large dynamic NOEs can be observed in macromolecules, the application of the 2D NOESY technique can provide accurate and reliable results. This makes NOESY a very powerful method for studying the structure of a macromolecule. For biological work it is particularly useful that NOESY spectra can be measured as easily in water solution as in deuterated solvents, so that the biological effects of the macromolecules are not altered or lost in nonprotonic or nonaqueous solvents.

Obviously, NOEs measured by the NOESY method, i.e., by the variants of basic pulse sequence (4.3-2), are not steady-state NOEs but instead are transient effects. Kumar et al. [246] have shown that the initial buildup rates of NOE are simply related to the internuclear distances. It is therefore mandatory for a quantitative analysis to measure NOESY spectra for several short d_m values ($d_m < 100$ ms when working with proteins of molecular weight in the vicinity of 20,000) in order to determine the buildup of the initial magnetization.

The precision of a NOESY measurement is affected by all the factors discussed above. In addition, spin diffusion can become quite efficient at high magnetic fields and can cause the NOEs to be less specific and hence less useful.

NOESY spectra must usually be combined with COSY spectra in structural determinations, but the time-consuming nature of the experiments is troublesome. Data acquisition times from 12 to 70 h are not unusual, and the requirement of maintaining identical conditions means that small drifts in temperature, resolution, and so on, may easily lead to mismatched spectra, especially in the case of complicated spectra containing hundreds of lines such as can be found in biological molecules [247]. A considerable amount of time can be saved if COSY and NOESY are measured simultaneously in a combined single experiment as proposed by Haasnoot et al. [247] (COCONOSY) and by Gurevich et al. [248] (COSY-NOESY). The experiment involves the extension of the normal COSY sequence with an extra 90° observation pulse and a separate data aquisition step. The idea of simultaneous measurement of two 2D NMR spectra is very attractive in terms of spectrometer time savings, although the instrumentation requirements make the extension of this principle to other 2D paired methods difficult.

Wagner [249] has proposed using "relayed NOESY" as a means of transferring NOE cross-peaks from crowded to sparsely populated spectral regions and thereby to identify connectivities that would otherwise be difficult to demonstrate because of degeneracy or overlap of resonances. One advantage, of course, would be to permit studies on larger molecules than those currently able to be studied.

Bremer et al. [244] have proposed a new method that produces skewed NOESY spectra and thus has been termed SKEWSY. The SKEWSY experiment involves a 180° pulse in the preparation period and is roughly similar to the sequence used in 2D exchange difference spectroscopy. The advantage of this method is that it requires no lineshape analysis. All information is thus obtainable directly from cross-peaks, and the crowded diagonal region can be avoided. A disadvantage is that although certain cross-peaks are enhanced, others (usually weaker ones) tend to be weakened.

It might also be noted that heteronuclear NOEs can also be measured by 2D techniques. The $^{13}C-^{1}H$ measurement known as HOESY [250, 251] is a typical example.

STRATEGY FOR USING 2D NMR SPECTROSCOPY

In the preceding sections we have treated and classified numerous 2D NMR methods according to the presence or absence of the mixing period in the pulse sequence. Although this division into correlated and resolved 2D spectra is important from the didactic point of view (since it is based on the physics underlying the experiments), it has little significance for the solution of problems by 2D NMR spectroscopy or for the choice of the method to solve a particular problem of structure determination. For such purposes we briefly review the methods discussed earlier, but this time we utilize a pragmatic point of view of the type of information that the various methods provide. Following this review, we address the general strategy for combining 2D methods to solve problems of structure determination or of complete spectral assignment.

5.1. SELECTION OF THE EXPERIMENT

The simplest type of problem for which a 2D method might be utilized is the extraction of a particular NMR parameter in the case when a conventional 1D spectrum is too crowded, overlapped, or simply not sufficiently resolved to yield directly the needed parameter.

Table 5.1 lists the NMR spectral parameters and the types of 2D NMR spectra from which the parameter can be extracted. The table can serve as only a rough guide for method selection since only basic experiments are listed; we have seen that modifications of the basic experiments have been designed to eliminate some features from the 2D spectrum. In practically all cases (except with NOE) there are a number of possible choices, and the proper choice of 2D method should include consideration of the required precision of the parameter as well as the time required for the experiment to be performed.

When the needed NMR parameter is a proton-coupling constant, J_{HH} (e.g., for a stereochemical study), precise values can be obtained by J-resolved methods. With limited data storage, basic correlation experiments usually provide less precise coupling constants values since the larger spectral

Table 5.1. NMR Parameters and Basic 2D NMR Experiments

NMR Parameter[a]	2D Experiment Yielding the Parameter[b]
J_{HH}	J-resolved (homonuclear)
	H,H-COSY
	H,C-COSY
	RELAY (homo- and heteronuclear)
J_{HC}	J-resolved (heteronuclear)
	H,C-COSY (undecoupled, FOCOUP)
J_{CC}	INADEQUATE
δ_H	J-resolved (homonuclear)
	H,H-COSY
	H,C-COSY
	RELAY (homo- and heteronuclear)
δ_C	J-resolved
	H,C-COSY
	RELAY (heteronuclear)
	INADEQUATE
NOE[c]	NOESY

[a] J stands for spin-spin coupling constant and δ for chemical shift. Subscript H denotes protons, but it can also represent any abundant NMR active nucleus. Subscript C denotes the carbon-13 nucleus, but it can also represent any other "rare" NMR active nucleus.
[b] Only basic experiments are considered. Some modifications eliminate the influence of a given parameter on the resulting 2D spectrum.
[c] Nuclear Overhauser enhancement.

width requires a higher sampling rate; thus the maximum achievable resolution is lower. (Of course, this limitation can be circumvented, if necessary, by using selective correlation methods or by using a variant in which the effect of proton shifts is eliminated.) On the other hand, heteronuclear correlations such as H,C-COSY or RELAY would be more useful in the case of a severely crowded proton spectrum. Then, the large spectral spread of carbon-13 chemical shifts can very effectively separate otherwise overlapping proton multiplets. Such heteronuclear correlations might also help in the case of systems with strong proton-proton coupling; the large one-bond $^{13}C-^1H$ coupling constants usually produce first-order ^{13}C satellite multiplets in a ^1H NMR spectrum. For these reasons J-resolved spectra are now seldom used even though homonuclear resolved methods have high sensitivity. If the multiplets are obscured by a strong singlet signal, multiple quantum filters can be used to eliminate the undesired singlet line.

Heteronuclear proton-carbon coupling constants are sometimes needed in studies of stereochemistry or bond hybridization. The discussion above about the precision of the measurement also applies to these coupling constants. However, the sensitivity of a H,C-COSY measurement without decoupling is lowered, but it still remains (thanks to the polarization transfer) somewhat higher than the sensitivity of the heteronuclear resolved method. The sensitivity of the heteronuclear resolved method can be increased, however, by using either a more efficient excitation scheme or by employing the so-called "reversed method" in which a ^{1}H NMR signal is detected only from those protons that are coupled to carbon-13 nuclei.

For the measurement of homonuclear carbon-carbon coupling constants (and isotopic effects) we have little choice, and it should be noted that for precise values a 1D INADEQUATE measurement is more appropriate than its 2D variant.

Chemical shift values that are needed for various purposes (e.g., substituent effect evaluation, chemical equilibria) are usually taken from appropriate correlation spectra. Very convenient are those 2D spectra (such as f_1-decoupled H,H-COSY or H,C-COSY measured with BIRD proton broadband decoupling) that will directly yield proton chemical shifts by means of spectral simplification without need of spectral analysis. Indirect detection of chemical shifts of rare nuclei is also advantageous because of enhanced sensitivity.

The sensitivity of these experiments varies to a large extent. To provide an approximate comparison, Morris [14] has compiled the approximate minimum time and the amount of material needed to perform a 2D experiment that produces an acceptable spectrum. His estimates are collected in Table 5.2. Also included for comparison are estimates for the simplest 1D NMR experiments. These estimates are based on performance experience with a 300-MHz spectrometer with a 250:1 signal-to-noise ratio for ^{13}C (standard ASTM test with 60% hexadeuteriobenzene in dioxane and 3.7-Hz line broadening) and 60:1 for ^{1}H (0.1% ethylbenzene, 1-Hz line broadening) in a 10-mm probe. A medium-sized organic molecule with a molecular weight of approximately 500 daltons and a T_1 of about 3 s for ^{13}C and 1 s for ^{1}H were assumed.

The actual demands for time and quantity of material needed to obtain satisfactory 2D spectra may exceed those given in the table by a considerable margin (e.g., by factor of 2 or 3) since the relaxation times might be longer, or since samples might have limited solubility, and so on. Nevertheless, the values in the table do provide a rough estimate for the relative demands of the experiments (e.g., if a 1D ^{1}H NMR spectrum requires n minutes time

**Table 5.2. Estimated Minimum Time and Amount of Material
for an NMR Experiment[a]**

NMR Experiment	Minimum Time[b] (min)	Minimum Amount[c] (mg)
1D ^{1}H NMR	0.1	0.005
1D ^{13}C NMR decoupled	0.1	5
H,H-COSY		
Low resolution	4	2
Normal resolution	30	5
High resolution	60	10
H,C-COSY		
One-bond	4	10
Long-range	6[d]	30
2D INADEQUATE	60	100
H,H,H-RELAY		
Low resolution	4	5
Normal resolution	30	10
H,H,C-RELAY	4	50
NOESY		
Low resolution	4	5[d]
Normal resolution	10	10
J-resolved		
Homonuclear	4	0.5
Heteronuclear	10	30

[a] Adapted from Morris [14].
[b] Minimum time needed to produce acceptable spectrum, assuming no limitations on sample
concentration; data processing not included.
[c] Amount of material needed to obtain adequate results from an overnight experiment
[d] Value not given by Morris [14].

averaging to provide an adequate signal-to-noise ratio, it will be necessary to allow at least 40 to 600 n minutes for measurement of a H,H-COSY spectrum).

When time requirements appear excessive, it is advisable to follow "the seven pillars of wisdom" (Section 2.5.2) for experiment sensitivity optimization. In some instances, however, extensive time averaging is dictated by phase cycling even though much less averaging would be sufficient for a good signal-to-noise ratio (e.g., in phase-sensitive H,H-COSY experiments or multiple quantum filtered correlations).

5.2.   STRATEGY FOR COMBINED USE OF 2D EXPERIMENTS

In the preceding section we were concerned with the simplest problem, the determination of one spectral parameter. Now we address briefly the other extreme, a full spectral assignment and structure determination in the case of a "completely" unknown compound.

No matter how powerful the combinations of NMR methods, NMR spectroscopy should not be taken as a handy substitute for elemental analysis, purity checking (e.g., by chromatography), molecular weight determination, and other methods of structure determination. The risk of drawing an erroneous conclusion from NMR spectra is high if the interpretation of the spectra is not backed by other methods. Therefore, the unknown compound under study should never really be "completely" unknown. Before attempting an interpretation of a spectrum, we should know whether or not the sample is pure, which elements (NMR inactive ones in particular) it contains, and so on. Usually, some clues to the possible structure of the sample are provided by its origin.

The decisive criterion for selecting the experiments or devising the strategy for solving the problem is the relative value or availability of spectrometer time. We shall consider two extreme cases: first, when spectrometer time is scarce; and second, when spectrometer time is plentiful and less expensive than the time of qualified personnel (the spectrometer operator and an analyst to interpret the spectra).

It would be a useless exercise to try to devise a universal and rigid strategy for solving a structural problem by a combination of 2D (and other) NMR experiments when spectrometer time is valued highly. Under such circumstances the operator and analysts must work together with versatility, flexibility, and imagination. The next experiment should be chosen and designed on the basis of all knowledge gained about the sample from previous experiments. A 2D experiment should be employed only when 1D NMR methods fail or become as time consuming as the 2D techniques. In many cases, 2D experiments can be specifically adapted to the problem at hand. Some examples of imaginative approaches are discussed by Morris [14] and by Bernstein [23] (e.g., phase cycling can be reduced when time averaging is not needed for sensitivity).

At the other extreme, when spectrometer time is not the decisive criterion, a heavier use of 2D experiments could be contemplated (e.g., to be run

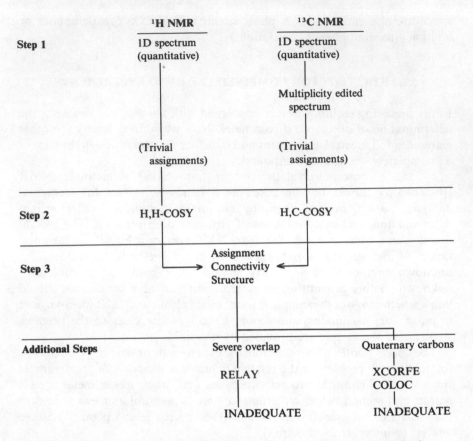

Figure 5.1 Chart of strategy for full spectral assignment and structure determination.

overnight) with a concomitant reduction of creative work by the analyst to a more or less routine exercise. (Full automation of the analysis, based perhaps on the symmetry of the spectra [252] and on pattern recognition [253–257], will no doubt occur in the near future). The demands on operator time are also drastically reduced since the experiments can be run unattended. Moreover, modern spectrometers equipped with a dual ^1H–^{13}C switchable probe allow "queuing" of a series of ^1H and ^{13}C NMR experiments. Under such circumstances, one generally applicable strategy would provide useful guidance to the experimentalist.

Powerful synergic combinations of 2D NMR experiments were recently reviewed by Bernstein [23]. Specifically designed strategies have been described for particular classes of compounds like small proteins (Wüthrich et al. [143, 144, 227–229, 232, 258] peptides (Kessler [26, 259]), or more

Table 5.3. Parameters and Mechanisms Detected in Some Correlation Experiments[a]

Experiment	Variant	f_1	f_2	Correlation	Mechanism
H,H-COSY	Basic	δ_H, J_{HH}	δ_H, J_{HH}	H↔H H	$^2J_{HH}, {}^3J_{HH}$
	f_1-decoupled	δ_H	δ_H, J_{HH}	H—C—C—C—	
	Long-range	δ_H, J_{HH}	δ_H, J_{HH}	H H H / —C—C—C—	$^nJ_{HH}, n \geqslant 3$
H,C-COSY	Basic	δ_H, J_{HH}	δ_C	H H H / C—C—C—C—	$^1J_{CH}$
	FUCOUP	δ_H, J_{HH}	δ_C, J_{CH}	H H H / C—C—C—C—	$^nJ_{CH}, n \geqslant 1$
	BIRD	δ_H	δ_C	H H H / C—C—C—C—	$^1J_{CH}$
	Long-range	δ_H, J_{HH}	δ_C	H H H / —C—C—C—C—	$^nJ_{CH}, n \geqslant 2$
	COLOC	δ_H, J_{HH}	δ_C	H H / —C—C—C	$^2J_{CH}, {}^3J_{CH}$
	XCORFE	δ_H, J_{HH}	δ_C	H H H / —C—C—C—	$^2J_{CH}, {}^3J_{CH}$
H,H,H-RELAY	Basic	δ_H, J_{HH}	δ_H, J_{HH}	H↔H↔H / —C—C—C—	$^2J_{HH}, {}^3J_{HH}$
H,H,C-RELAY	Basic	δ_H, J_{HH}	δ_C	H↔H H / C—C—C—C—	$^3J_{HH}, {}^1J_{CH}$
INADEQUATE	Basic	$\delta_C + \delta_C$	δ_C, J_{CC}	H H H / C—C C	J_{CC}
	& INEPT or DEPT	$\delta_C + \delta_C$	δ_C, J_{CC}	H H H / C—C—C—	$^1J_{CH}, J_{CC}$
NOESY	Basic	δ_H, J_{HH}	δ_H, J_{HH}	H↔H / —C C—	NOE (homo.)
HOESY	Basic	δ_H, J_{HH}	δ_C	H H / —C C—	NOE (hetero.)

[a] Detected nuclei are indicated by underlining, e.g., C̲.

generally, oligomers [23] and aromatic compounds [23]. The specialized strategies utilize particular features of such classes of compounds. For example, biological oligomers (including peptides, oligosaccharides, nucleosides, etc.) contain distinct monomeric units with spin-spin couplings within the units but with little or no coupling between them. Also, large NOE values in these molecules make the results of NOESY experiments diagnostically valuable.

A general useful strategy for total spectral assignment and structure determination of an "unknown" organic compound is comprised of the steps delineated in Figure 5.1 (modified general strategy 2 of Bernstein [23]).

Although 2D experiments can be run "blind," they are much more efficient if their measuring parameters are adjusted according to the 1D spectra (step 1). In many instances, it is advantageous to run 1D spectra (also) for quantitative results so that intensity ratio information can be utilized later and so that mixtures can be noticed at an early stage. Interpretation at this step 1 level can usually provide only trivial assignments (and identification of obvious molecular fragments) unless a considerable amount of information about the compound is known from other sources.

The 2D COSY experiments in step 2 are usually run unattended. If time permits, some of the additional experiments anticipated in step 4 can be queued for the same overnight run. Subsequent interpretation is greatly facilitated if all 2D spectra are run under such conditions that the spectra are easily juxtapositioned. Specifically, the proton decoupler frequency should be the same as the observing frequency in H,H-COSY spectra, and the spectral widths (along the f_1 axis) should also be the same. Interpretation of each of the COSY spectra has to be carried out along the lines described in Chapter 4, but combined simultaneous interpretation of the two spectra is very efficient.

When the problem cannot be solved in step 3, some additional experiments must be performed. The additional experiments must be chosen according to the type of missing information or correlation. Some guidance is provided by Table 5.3, which lists some of the most useful 2D correlation experiments in an approximate descending order of usefulness for analysis of medium-sized molecules. Other experiments are, however, also possible. In interpretation of the results (especially of the additional experiments), one must be careful since most of the conclusions are based on the relative magnitudes of the long-range couplings. The coupling constants do not always have the expected relative magnitudes.

APPENDIX

A.1.   PHYSICS OF PULSED NMR AND THE VECTOR MODEL

Let us think about what happens with a nuclear spin system during the course of a one-dimensional NMR experiment such as the one depicted schematically in Fig. 2.1. An explanation using a simple example will help us brush up on some terms that are used frequently throughout this book.

When we interpret an NMR spectrum (i.e., when we assign the chemical shifts, coupling constants, etc.), we usually view the spectrum as if it had originated from one single molecule that has all chemical shift and coupling constant values. It is not possible, however, to understand what is happening during even the simplest NMR measurement with this kind of an approach. An NMR spectrometer measures the macroscopic behavior of the entire sample, which consists of a large number of small quantum systems (molecules, atomic nuclei, etc.). For the sake of illustration, let us consider the measurement of the ^{1}H NMR spectrum of chloroform, ^{1}H^{12}CCl$_3$. We will measure the spectrum on a spectrometer operating at 100 MHz (i.e., in a magnetic field $B_0 = 2.349$ T, where T is the tesla, the unit of magnetic induction). If the volume of the HCCl$_3$ sample in the NMR tube is 1 mL, the sample contains about 10^{21} protons. The behavior of each and every one of the 10^{21} protons in the sample is controlled by the laws of quantum physics. The measurable macroscopic quantities (observables) are the vector sums of corresponding microscopic quantities of each of the 10^{21} protons in the sample.

Before we place our sample into the magnet of the NMR spectrometer, the nuclear spins of the chloroform protons have random orientations (i.e., no direction is preferred). The vector sum of all 10^{21} individual nuclear magnetic moments is therefore zero. After we have placed the sample into the strong static magnetic field B_0 of the spectrometer, the nuclear spins are no longer randomly oriented. According to quantum physics, each of the proton spins can have only one of two allowed orientations with respect to the magnetic field. (In general, there are $2I + 1$ allowed orientations for nuclei with spin I; protons have a nuclear spin $I = \frac{1}{2}$). The two nuclear spin orientations (the two "states") have different energies; therefore, two energy levels exist. After a period of time characterized by the spin-lattice relaxation time T_1, an equilibrium distribution of proton spins in the two energy levels is achieved. At equilibrium, there are more protons in the lower energy level, in which each proton has its nuclear magnetic moment $\bar{\mu}$ oriented essentially along the direction of the magnetic field B_0. These protons all have a magnetic quantum number $m = +\frac{1}{2}$; they are in the $<+\frac{1}{2}>$ or $<+>$ state. The number of protons in the higher energy level is smaller, and each of these protons has its nuclear magnetic moment $\bar{\mu}$ oriented essentially against the direction of the external magnetic field B_0. The magnetic quantum number of these latter protons is $m = -\frac{1}{2}$; they are in the $<-\frac{1}{2}>$ or $<->$ state. (A macroscopic magnet such as the needle of a compass will align itself exactly along the direction of the magnetic field. Nuclear spins are said to be oriented "parallel" or "antiparallel" with the magnetic field when the projection of their magnetic moments in the direction of the magnetic field is parallel or antiparallel with that field.)

According to Boltzmann distribution, the energetically more favorable (lower) level will be populated by only 10^7 more protons than there will be in the higher energy level. Remember, there are about 10^{21} protons altogether in the sample, and for equal population of the two levels we would therefore have 5×10^{20} protons in each of them. Thus the difference in the populations, 1×10^7, is indeed very small in comparison with the populations in the two levels. It is this very small population difference that is the source of the signal that a NMR spectrometer detects.

As a consequence of nuclei having both a magnetic and an angular momentum, the magnetic moment vectors $\bar{\mu}$ of all protons in the sample precess about the direction of the magnetic field B_0 with the Larmor precession frequency

$$f_0 = \gamma B_0 \qquad (A.1\text{-}1)$$

(where γ is the magnetogyric ratio divided by 2π). All moments precess with the same frequency. When a 100-MHz NMR spectrometer ($B_0 = 2.349$ T) is used, the Larmor frequency of protons is 100 MHz. The relative orientation of the moments, however, is random, and the vectors cover the surface of the cones shown in Fig. A.1 in a uniform way. This type of motion, when the vectors have arbitrary phases, is described as "incoherent." We shall encounter other situations in which some vectors will move with the same phase or with a fixed mutual orientation, and then we refer to "coherent" types of motion.

Since there is an excess of protons in the lower energy level ($m = +\frac{1}{2}$), and since their motion is incoherent, the vector sum of all 10^{21} nuclear magnetic

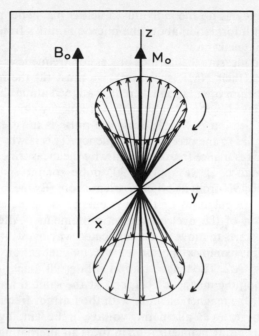

Figure A.1 Magnetic moments of nuclei and their vector sum; nuclear magnetization M_0 at equilibrium (schematically); magnetic moments parallel with external field $\bar{B}_0$ belong to spin state $<+\frac{1}{2}>$; those with the opposite orientation belong to spin state $<-\frac{1}{2}>$; the heads of the magnetic moment vectors move on the dashed circles during precession with Larmor frequency f_0.

moments, the macroscopic nuclear magnetization $\bar{M}$, is exactly parallel at equilibrium ($\bar{M}_0$) to the applied external magnetic field $\bar{B}_0$. The magnetization M_0 is proportional to the population excess of protons in the lower energy level. (In NMR experiments we can achieve a different orientation of the magnetization by an appropriate combination of rf pulses. For example, when the magnetization has an orientation opposite to that just described, it means that the populations of the energy levels are inverted; that is, that the higher energy level is more populated than the lower one.)

Although each of the individual nuclear moments behaves according to the laws of quantum mechanics, their sum behaves as a classical vector that can assume an arbitrary orientation in space. The magnetization, however, moves in a magnetic field in a way similar to the constituent nuclear magnets of which it is composed. When the magnetization is not at its equilibrium position, it precesses about the direction of the magnetic field with the same

Larmor frequency as do the individual nuclear magnetic moments. This is why we can get information about the microquantities from measuring this macroscopic magnetization.

To be able to measure the Larmor precession frequencies of the protons in the sample (i.e., their NMR spectrum), we must tilt the magnetization $\bar{M}$ from its equilibrium orientation where it is aligned along the magnetic field. This tilt is achieved by an rf pulse with a frequency close to the precession frequency (see Section A.4.1). In fact, the pulse is most effective when its frequency is equal to the precession frequency. This is why the frequency is often called the resonance frequency and why the measuring method is called magnetic resonance. If we apply a 90° pulse to our sample, we tilt the magnetization by 90° from its original orientation. (By means of this pulse we have made the populations of the two energy levels equal and created a coherent motion of the nuclear magnetic moments.) After the pulse, the magnetization starts to move in a complicated way back into its equilibrium orientation. This complex motion of the magnetization is schematically depicted in Fig. A.2. The magnetization rotates 90° back to its equilibrium orientation along the magnetic field, and at the same time, it rotates about the direction of the magnetic field $\bar{B}_0$ with the Larmor frequency f_0. Just as a magnetic rotor creates an alternating voltage in the windings of a stator in a generator, the rotating magnetization induces an alternating voltage in a coil mounted around the sample inside the magnet of the spectrometer. The voltage induced is our NMR signal. The time dependence of the voltage [or the NMR signal $s(t)$] induced by the nuclear magnetization in the (receiver) coil and amplified in the receiver of our spectrometer is shown in Fig. A.3. The time dependence of the induced signal is generally denoted as a FID (free induction decay).

The smooth exponential decay of the amplitude of the damped oscillations in Fig. A.3 is characterized by the spin-lattice relaxation time T_2, which describes the loss of coherence among the nuclear magnetic moments. In our simple example, the frequency of the oscillations is the Larmor precession frequency of the protons in chloroform. The frequency (or the spectrum) can be easily determined from this time dependence simply by counting the number of oscillations per second.

A sample containing different types of protons (i.e., protons with different chemical shifts, which have different Larmor frequencies) produces a much more complex FID (see the FID in Fig. 2.1). The FID in such a case is a superpositioning (or interferogram) of all the simple FIDs, each of which is due to one type of proton, with one particular frequency (if we assume no spin-spin coupling) and one particular spin-spin relaxation time. It is not possible to determine the frequencies of the Larmor precession of all the different types of protons present in the sample directly from this complex

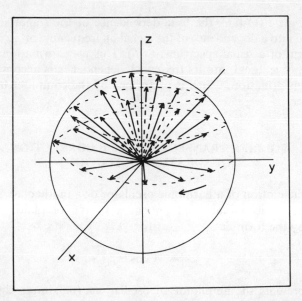

Figure A.2 Return of the magnetization M to equilibrium after a 90° pulse (schematic); spin-spin relaxation neglected).

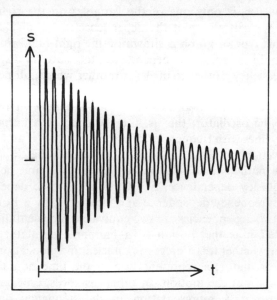

Figure A.3 Simple FID; the dependence of the NMR signal, s, on time, t, $s(t)$. The usual time scale is in the range of microseconds to a few seconds; the signal can be measured in voltage units.

FID. First, the FID [i.e., the time dependence of the signal $s(t)$] must be transformed into a dependence of the signal on frequency, or, in other words, into the form of a usual spectrum, $S(f)$. This transformation of the time dependence of a signal into its frequency dependence is accomplished by a Fourier transformation, which is performed by the computer built into the spectrometer.

A.2.   FOURIER TRANSFORM, PHASE CORRECTION, AND QUADRATURE DETECTION

We know the motion of a harmonic oscillator or a mathematical pendulum from physics. The actual displacement x of the oscillator at time t is expressed by the formula

$$x = A \cos 2\pi f_0 t \tag{A.2-1}$$

were A represents the amplitude of oscillations (maximum displacement), and f_0 is the frequency of oscillations (in hertz). The angle $2\pi f_0 t$ is known as the phase. Most textbooks give only the time dependence of displacement, as shown in the graph on the left-hand side of Fig. A.4a.

Time, however, is only one of the variables on the right-hand side of Eq. (A.2-1). We can also plot the dependence of the displacement on the frequency; this type of graph is shown on the right-hand side of Fig. A.4a. According to this less familiar dependence, there are no displacements that vary with frequency f other than $\pm f_0$ (in other words, all periodic motions except the ones with frequencies $\pm f_0$ have an amplitude of zero). In spectroscopic terms the dependence of displacement on frequency is the spectrum of the oscillator; that is, the spectrum of a harmonic oscillator contains only lines with frequencies of $+f_0$ or $-f_0$ (the $+$ and $-$ signs denote opposite directions or senses of rotation, which will be elaborated later).

There is a mathematical relationship between the time dependence, $s(t)$, and the frequency dependence, $S(f)$. From the time dependence we can calculate the frequency dependence and vice versa, by a Fourier transform (FT). These two dependencies are two equivalent representations of the same event, whether it is the motion of a harmonic oscillator or a swinging pendulum, or whether it is a precessing nuclear magnetization. In the case of a harmonic oscillator, a representation in the time domain gives a very concrete idea about the motion. In other instances, especially in the more complicated ones, a representation in the frequency domain is more convenient. The latter representation reduces the rather complicated time dependence into simple single-frequency components. A Fourier transform

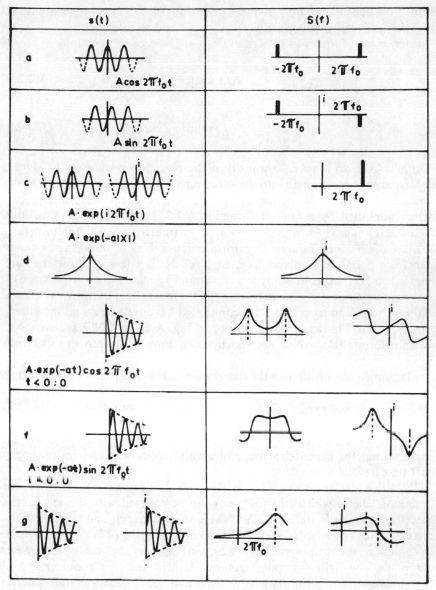

Figure A.4 Pairs of functions related by Fourier transforms according to Chempeney [260].

159

of the time dependence into frequency dependence is a mathematical means of making the dependence simpler to "read" and to interpret.

Fourier transforms can be written in many forms, for example,

$$S(f) = \int_{-\infty}^{+\infty} s(t) \cdot \exp(-i2\pi ft)dt \qquad (A.2\text{-}2a)$$

$$s(t) = \int_{-\infty}^{+\infty} S(f) \cdot \exp(-i2\pi ft)df \qquad (A.2\text{-}2b)$$

Formula (A.2-2a) is used to transform the time dependence of the signal, $s(t)$, from the time domain into the spectrum (frequency dependence), $S(f)$, which is of course in the frequency domain. The reverse transformation can be accomplished by means of formula (A.2-2b). Mathematical functions such as $s(t)$ and $S(f)$, which are related by transforms (A.2-2a) and (A.2-2b), are often called Fourier transform pairs. Tables of important Fourier transform pairs are available (see, e.g., ref. 260); a few of those that are relevant to NMR spectroscopy are shown in Fig. A.4. Note the presence of the imaginary unit $i(\sqrt{-1})$ in the expressions for the FT; transforms of some real functions lead to complex functions that have both real and imaginary parts (e.g., see Fourier pairs b, d, and f in Fig. A.4). In NMR spectroscopy we transform FIDs, which are functions of time, $s(t)$, into the spectrum $S(f)$.

The simple ideal FID (like the one shown in Fig. A.3) can be expressed as

$$s = A \exp(-t/T_2) \cos(2\pi f_0 t + \phi) \qquad \text{with } \phi = 0 \qquad (A.2\text{-}3)$$

According to Fig. A.4e, the Fourier transform, $S(f)$, of the FID has a real part that has the Lorentzian shape of an absorption line and an imaginary part that has the shape of a dispersion line.

By a Fourier transformation of this ideal FID, we obtain a pair of functions, one of which (the real one) can be equated with the absorption spectrum (a v-mode signal in CW NMR spectroscopy), and the other (the imaginary) can be equated with the dispersion component (a u-mode signal). In practice, the transform is performed digitally by the spectrometer computer. The digital computer cannot deal with the continuous analytical functions assumed in Eq. (A.2-2), but the function is converted into a series of discrete values that must satisfy certain requirements that will be noted later. Also, the result of the transform is given as two series of discrete values. One series represents the real part of the FT, and the other series, the imaginary part. The two series are stored in the computer in separate locations. As you might guess, an FID produced by the measurement of even the simplest sample will not be an ideal FID such as that shown in Fig. A.3. In the simplest case, the real FID satisfies expression (A.2-3), but the angle ϕ is not zero (i.e., the beginning of the FID is shifted). In this case the real and imaginary parts, $\text{Re}(S(f))$ and $\text{Im}(S(f))$, of the transformed function no longer have the shape of pure absorption and dispersion lines; the two modes are intermixed. However, the pure shapes can be obtained from the two parts as their linear combination:

$$\sin\phi \cdot \text{Re}(S(f)) + \cos\phi \cdot \text{Im}(S(f)) \qquad \text{(A.2-4)}$$

where the angle ϕ is usually a linear function of frequency (because of the frequency dependence of the spectrometer performance), that is,

$$\phi = a + bf \qquad \text{(A.2-5)}$$

The values of the constants can be adjusted by the operator by means of a trial-and-error procedure in which the operator seeks a pure absorption lineshape in the spectrum. The process is known as phase correction of the spectrum. Naturally, there are computer programs that perform phase correction very effectively.

The simple cosine wave $s(t)$ of Fig. A.4a has in its spectrum $S(f)$ two lines with frequencies $+f_0$ and $-f_0$ (see the discussion above). If we recall how a cosine dependence can originate from a periodic circular motion, the origin of the two signs becomes clear. As shown in Fig. A.5, the cosine dependence is obtained from a projection of the circular motion onto the x axis.

Vectors A^- and A^+, rotating clockwise and counterclockwise, respectively, can represent magnetization components M^S or M^I or radiofrequency field B_1 components. Obviously, the same cosine dependence is obtained from clockwise (vector A^-, dashed line) and counterclockwise (vector A^+, solid line) rotations. That is why the FT gives the two possible signs for the frequencies; it cannot distinguish the two opposite directions of rotation. When we measure the time dependence of a NMR signal, the FID obtained is proportional to a projection of the rotating magnetization vector onto one of the fixed or laboratory framework axes, and naturally, FT produces two lines with opposite frequencies. Although the magnetization rotates in one particular direction, an FT of a single FID does not permit a determination of this direction. (Some spectrometer software, however, is arranged so that you can see only the lines with frequencies of one sign.) If we project the circular motion at the same time onto the other axis, the y axis, the projection will have the phase dependent on the direction of the rotation (see the right-hand portion of Fig. A.5). The two projections are the two components of

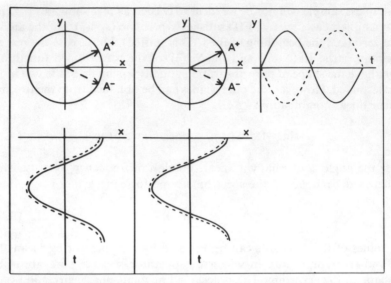

Figure A.5 Periodic circular motion of two vectors in opposite directions, and their projections onto coordinate axes: left; projection onto the x axis, single detection; right, projections onto the x and y axes, quadrature detection.

the vector that completely identify the vector in a plane; no ambiguity remains regarding its phase. It is sufficient to equate the projection on the x axis to $\text{Re}(s(t))$, and the projection on the y axis to $\text{Im}(s(t))$, which are two parts of the same complex function $s(t)$, and then tc perform a Fourier transform of this complex function. The FT (Fig. A.4c) gives only one line with the frequency sign corresponding to the direction of rotation.

In NMR spectroscopy we achieve this sign discrimination by quadrature detection. Essentially, quadrature detection during FID acquisition consists of recording the time dependences of the two projections of the nuclear magnetization motion onto two perpendicular axes. There are various technical means whereby quadrature detection can be implemented on different spectrometers (for a review, see, e.g., ref. 5). The most obvious method is the detection of the NMR signal by two measuring channels, A and B, with one channel measuring the x component and the other channel measuring the y component. During the subsequent FT, one of the FIDs is assumed to be the real and the other FID to be the imaginary part of the complex FID that has to be transformed. Quadrature detection has many advantages for NMR spectroscopy [262, 263], in general, and the advantages are accentuated in 2D NMR. Quadrature detection permits us to set the transmitter frequency in the center of the spectrum (since the FT will differentiate the positive and negative frequency differences), and thus computer storage requirements are considerably reduced and transmitter power is more effectively utilized.

Obviously, this principle of quadrature detection can be used in 2D NMR spectroscopy only within the detection period during FID acquisition, or, stated somewhat differently, only along the f_2 axis. Analogous methods of achieving frequency sign discrimination along the f_1 axis (i.e., during the evolution period) are more complicated (for details, see Section A.7).

In 2D NMR spectroscopy there are always complex interferograms available for the second FT as the outcome of the first FT. If the evolution of the spin system during time t_1 affects only the amplitude of the interferograms (amplitude modulation), the second FT produces spectra $S(f_1)$ symmetrical around $f_1 = 0$, and it will not distinguish positive and negative frequencies in the f_1 dimension. Similarly, as quadrature detection encodes the sign of rotation during the detection period into the phase of the FIDs, information about the sign of rotation during the evolution period must be encoded into the phase of the interferogram (combined phase and amplitude modulation). The type of modulation produced by a 2D experiment is determined by the pulse sequence and phase cycling scheme. Of two otherwise identical experiments, the one that produces a phase-modulated interferogram is usually advantageous since it permits the use of two to four times weaker decoupling, reduces computer storage requirements by half, and permits the achievement of the same resolution with only half the number of t_1 increments [97] (Sections 2.5 and A.7).

A.3. ROTATING FRAME OF REFERENCE. SINGLE rf PULSE. SIGNAL DETECTION, AND PHASE CYCLING

Initially, the concept of a rotating frame of reference (or rotating frame) might appear as an unnecessarily complex notion, but its introduction will soon pay off. In NMR spectroscopy we measure the motion of nuclear magnetization, and the description of the complicated motion of nuclear magnetization is considerably simplified if we employ the concept of a rotating frame. Then, the motion can usually be more easily visualized (and also more easily treated by theoreticians), and the experiments thus become easier to understand.

The standard "laboratory frame of reference" is a set of rectangular Cartesian axes (x, y, z), which form a reference frame that is fixed in laboratory space and can be visualized by looking into a corner of the laboratory [5]. The two intersections of walls and floor form the x and y axes, and the intersection of the two walls forms the z axis. This fixed reference will be hereinafter termed the "fixed (or laboratory) frame." The axes will be placed so that the magnetic field $\bar{B}_0$ in a NMR experiment will lie along the direction of the z axis. (Some authors choose to place the direction of the magnetic field along the direction of the $-z$ axis.)

The rotating frame, which will be designated (x', y', z'), is also composed of a rectangular set of Cartesian axes with the z' axis parallel to the z axis of the fixed or laboratory frame, but the x' and y' axes rotate around the laboratory z axis with the frequency f (hertz). If you wish, you can imagine the rotating frame as a frame of reference attached to a revolving turntable of a record player [5].

We have already noted that if the magnetization vector $\bar{M}$ is not parallel to the external magnetic field $\bar{B}_0$, it precesses around the laboratory z axis with the Larmor frequency f_0 (hertz). Now, if we choose frequency f of the rotating frame to be equal to the Larmor frequency f_0 (with the same direction of rotation), the frame of reference rotates with the magnetization and the magnetization is stationary relative to the rotating frame. (In an analogous fashion, the label on a revolving record would appear to us to be stationary if we were to climb onto the rotating turntable.) The complicated motion of magnetization (Fig. A.2) after a pulse is thus simplified, and the rotation has disappeared in the rotating frame. From the viewpoint of an observer located on the rotating frame, the magnetization behaves as if it were not subject to the external magnetic field B_0.

The rotating frame permits a simple description of the effect of an rf pulse on the nuclear magnetization. The rf field is a magnetic field that oscillates with frequency f_r in the rf range (megahertz). An rf field with amplitude $2B_1$ is created in the NMR probe by a coil. The voltage for the coil is provided by an amplifier driven by the transmitter from either the observing or the decoupling channel. The oscillating magnetic field lies in the direction of the laboratory x axis (i.e., in the direction that we have deliberately selected to be the x axis of the fixed laboratory frame). We can regard the oscillating field to be composed of two rotating fields (within the fixed laboratory frame), and both have the same amplitude B_1 but rotate in opposite directions with frequencies f_r and $-f_r$. (The two B_1 fields correspond to vectors A^+ and A^- in Fig. A.5, and the oscillating field is the vector sum of the two fields.) The effect of field B_1, which rotates in the opposite direction to the precession of magnetization can be neglected, since the difference in the two frequencies (f_0 and f_r) is too large. In this case we choose the rotating frame associated with the rotating magnetic field B_1, which has the same direction of rotation as does the magnetization. That is, we select a rotating frame with a rotating frequency $f = f_r$. In this framework, magnetic field B_1 is stationary, and we choose the direction of field B_1 for the direction of the x' axis of the rotating frame.

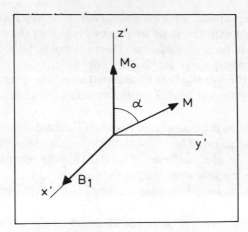

Figure A.6 Rotation of magnetization $\bar{M}$ under the influence of an rf field $\bar{B}_1$ in a rotating frame.

At resonance the Larmor frequency f_0 is equal to the transmitter frequency, f_r, which is also the frequency with which our rotating frame rotates. Provided that we start from equilibrium, the nuclear magnetization vector $\bar{M}$ points in the direction of the z' axis. When we switch on the rf field, the magnetic field $\bar{B}_1$ is perpendicular to the magnetization $\bar{M}$, and its direction is the direction of the x' axis. Under the influence of this rf field, the magnetization rotates (in the rotating frame) around the $-x'$ axis with a frequency

$$f_1 = \gamma B_1 \qquad (A.3\text{-}1)$$

(See Fig. A.6; this rotation is the Larmor precession in field B_1.) When the rf field is switched on for the duration of time t (the pulse width or length, t, is usually in the microsecond range), the magnetization turns by an angle

$$\alpha = 2\pi\gamma B_1 t \qquad (A.3\text{-}2)$$

The rotated angle determined by Eq. (A.3-2) is known as the flip angle. We control the flip angle by selecting the pulse length t and the rf field amplitude (the rf power on a NMR spectrometer console). Pulses with a 90° flip angle are referred to as 90° or $\pi/2$ pulses, and pulses with 180°, 270°, or 360° flip angles are given analogous names (for pulse calibration, see Section A.10).

To understand the more complicated pulse sequences that consist of several pulses, we should be aware that our rotating frame of reference continues to rotate with the same frequency even after the pulse is complete and the rf field has been switched off. The next pulse, although it may have the same frequency f_r and amplitude $2B_1$ as the first pulse, does not necessarily have its magnetic field $\bar{B}_1$ oriented along the x' axis of the rotating frame. The orientation of field $\bar{B}_1$ in the rotating frame is controlled by the transmitter phase.

A change of 90° in the transmitter phase shifts the direction of the $\bar{B}_1$ field to the y' axis, and so on. The transmitter phase (or pulse phase) thus has a significant effect on the outcome of an NMR experiment. This is amply demonstrated by the following simple experiment.

Suppose that we have a simple sequence of two 90° pulses, the second one following immediately after the first. The first pulse rotates the magnetization from its equilibrium orientation to the direction of the y' axis. If the second pulse has the same phase as the first one, its $\bar{B}_1$ field also falls in the direction of the x' axis, and the pulse rotates the magnetization from the y' axis to the $-z'$ axis. The total effect is a 180° rotation of the magnetization. If the phase of the second pulse is shifted by 180° with respect to the phase of the first pulse, the magnetic field $\bar{B}_1$ points during the second pulse in the direction of the $-x'$ axis, and the magnetization rotates back to its equilibrium direction along the z' axis. The results of the two experiments are quite different. For this reason, the pulse phase has to be added to the flip angle in order to characterize the pulse. For the sake of brevity, the pulse phase is usually given as the axis of the rotating frame along which the $\bar{B}_1$ field is oriented during the pulse (but the prime is omitted). Therefore, we have a $90°_y$ pulse or $90°_{-y}$ pulse. Since the choice of the rotating frame is arbitrary, the information on the pulse phases has the relative meaning that it relates phases in different pulses of the sequence (see Appendix Section A.8).

When the frequency of the rotating frame f_r, which is equal to the frequency of the transmitter and receiver, is not equal to the frequency of magnetization f_0, the magnetization rotates in the rotating frame with a difference frequency $f_0 - f_r$. It behaves as if it were exposed to the magnetic field

$$B_{ef} = B_0 - f_r/\gamma \qquad (A.3\text{-}3)$$

In such a case the transformation into the rotating frame does not completely remove the rotational motion of the magnetization, but its frequency is reduced from megahertz (in the fixed laboratory frame) to a few kilohertz (in the rotating frame). This is the standard situation in samples with several nonequivalent nuclei that have different chemical shifts. The frequency difference $f_0 - f_r$ is the chemical shift (in hertz) of the corresponding nuclei relative to the frequency of the transmitter.

There is a close relationship between the rotating frame and FID detection. The precessing magnetization induces an alternating voltage in the receiver coil in the NMR probe. The frequency of the alternating voltage is the frequency of precession f_0. Usually, the voltage is amplified in a preamplifier and in the receiver. For a number of technical reasons [262] the signal is subjected to synchronous detection which juxtaposes it with a reference signal of frequency f_r derived from the transmitter. The signal after synchronous detection has its frequency reduced by f_r. The signal has the same frequency that the magnetization has in the rotating frame. (For example, if the transmitter frequency equals the magnetization frequency in the fixed laboratory frame, the magnetization is stationary in the rotating frame and the FID detected has the shape of a simple exponential decay without oscillations. The difference frequency is therefore zero.) The shape of the signal after synchronous detection depends on the phase of the reference signal. By a suitable choice of the phase of the reference signal, we can prevent the detected signal from corresponding with the voltage induced in the receiver coil mounted along the y' axis but to correspond instead to the signal which would be detected in a coil mounted along the x' axis (where there is no receiver coil).

Many 2D NMR experiments require systematic changes in the phases of transmitters in both the observing and decoupling channels and also in the receiver phase. In general, we speak about phase cycling as the recipe for phase alternation necessarily involves some repetition after a certain number of passes through the pulse sequence. We should note that simple phase cycling is almost always used in 1D NMR spectroscopy in order to reduce noise of some type.

The well-known advantage of pulsed FT NMR spectroscopy is the possibility of repeating the measurement with a high repetition rate and thus improving the signal-to-noise ratio by adding the FIDs as they are recorded successively. During this series of additions (often called time averaging), the signal intensity increases faster than the incoherent noise, which appears to be averaged out. Coherent noise can also be reduced by a suitable cycling of the transmitter and receiver phases. In the case of single detection (as contrasted to quadrature detection), it is sufficient to change the transmitter phase in steps of 180° in two consecutive passes through the pulse sequence and alternately add and subtract the FIDs detected. Within the rotating framework, the 180° change of transmitter phase corresponds to a change of orientation of field $\bar{B}_1$ from its original direction pointing along the x' axis so that it now points in the $-x'$ direction. Consequently, the magnetization is

Table A.1. Phase Cycling for Quadrature Detection

Scan Number	Transmitter Phase	$\mathrm{Re}(s(t))$	$\mathrm{Im}(s(t))$
1	0°	x	y
2	90°	y	$-x$
3	180°	$-x$	$-y$
4	270°	y	x

no longer rotated from the equilibrium orientation into the direction of the y' axis but instead is now oriented in the $-y'$ direction, which corresponds to a multiplication of the odd FID signals by (-1). To sum constructively the FID signal multiplied by (-1), we have to subtract this FID from those that were not multiplied by (-1), that is, from those obtained in even scans. The data handling described is equivalent to changing the receiver phase synchronously with the transmitter phase.

In the case of quadrature detection, the phase cycling is more complex. The transmitter phase is incremented by 90° in each pass, and the x and y components (or signals in channels A and B) are added to the real (Re) and imaginary (Im) part of the complex signal $s(t)$ according to Table A.1.

Again, this data routing is equivalent to synchronously incrementing both the transmitter and the receiver phase by 90° in each successive pass. The scheme is known as CYCLically Ordered Phase Sequence (CYCLOPS) [263]. When the total number of accumulations is a multiple of 4 (to ensure that an integral number of phase cycles is accomplished), CYCLOPS corrects for several types of hardware errors [263, 264]. (Some spectrometers, however, require that the total number of accumulations be a multiple of eight for complete artifact suppression.)

In heteronuclear NMR pulse experiments, which include pulses that affect different atomic nuclei, it is customary to employ the notion of a doubly rotating frame. It is difficult to imagine a frame of reference that rotates with one frequency for nuclei of one type (e.g., for ^{1}H), and with another frequency for nuclei of a second type (e.g., ^{13}C). Nevertheless, the concept is useful when the resonating frequencies of the two types of nuclei are markedly different. Then, the nuclei or their magnetizations feel only those rf pulses that have a frequency close to their Larmor frequency (i.e., in our example, ^{1}H nuclei feel pulses with a frequency of 100 MHz, while ^{13}C nuclei are affected only if the frequency is close to 25 MHz). We can split the overall nuclear magnetization into the magnetizations of individual types of nuclei. The magnetization of nuclei of one type is stationary only in the rotating frame that rotates with the precession frequency of this magnetization. Also, the pulse phase (i.e., the orientation of field $\bar{B}_1$), is meaningful only in a framework rotating with the appropriate frequency.

A.4. ELEMENTS OF THE PULSE SEQUENCE AND THE VECTOR MODEL

The considerable number of published NMR pulse sequences (see the review by Turner [17]) has been constructed with the use of only a rather small number of the building blocks that we call pulse sequence elements. When the pulse sequence is being programmed, the elements are usually represented by separate commands or instructions within an instruction set of the sophisticated pulse programmer contained in a modern spectrometer.

In addition to various pulses of observing and decoupling frequencies, the pulse sequence contains different time delays, evolution and detection periods, continuous irradiation or magnetic field gradient pulses, and so on. Physical events (e.g., a laser pulse that initiates a chemical reaction) could also be used. These elements can be employed in any order and with various characteristics such as duration time of the element, frequency, and phases and their cycling. As a result, we have an enormous number of possible combinations, many of which can lead to useful NMR pulse sequences. When we want to reproduce a pulsed NMR experiment, we must know all of the pertinent details about the pulse sequence, that is, the order of the elements within the pulse sequence and all their characteristics. These data are usually provided in a scheme of a pulse sequence, which might be accompanied by a table or a formula giving the details of the phase cycling. The pulse sequence scheme is the timing diagram of the pulse sequence.

There are two customary means of presentation of pulse sequence schemes. The first one, which is used almost exclusively in this book, is graphical (see any pulse sequence scheme in the figures given in the main body of the text). The graphical presentation certainly gives a clear picture of the order of events, but it consumes considerable space. For this reason, journal editors prefer a second means of presentation, a linear presentation. A linear pulse sequence scheme can be printed directly in a horizontal row of text. For example, the pulse sequence from Fig. 2.5a can be written in a linear form as

$$90°(I) - t_1 - 90°(I), 90°(S) - \text{detection (S)}$$

In the remaining part of this appendix we explain the effects of pulse sequence elements on spin systems and on measurement results. We use the vector model that describes the simplest heteronuclear spin system "IS," for which we use $^1H^{13}CCl_3$ as an example. Extrapolation to slightly more complex spin systems such as I_2S and I_3S is not difficult. According to the vector model, we shall not be interested in the motion of the magnetic moments of individual nuclei, but we shall instead concentrate on the observable macroscopic magnetization. Of course, we already know that the magnetization is a vector sum of the individual nuclear magnetic moments, and that its component in the direction of magnetic field $\bar{B}_0$ (i.e., the M_z component) is proportional to the difference in energy-level populations. Furthermore, we also know that the transversed component in the x, y plane (i.e., M_{xy}) measures the degree of coherence achieved among the precessing nuclear magnetic moments.

Our simplest heteronuclear spin system consists of two nuclei, I and S, each of which has spin $\frac{1}{2}$. The spin-spin interaction J_{IS} (in hertz) is weak; it is much smaller than the difference of resonance frequencies $f_I - f_S$. (In the heteronuclear spin system the difference can be several tens or hundreds of megahertz; in a homonuclear spin system, however, the difference corresponds to the difference in chemical shifts, which can vary from 0 to about 10^5 Hz.)

The entire NMR spectrum of the IS spin system contains four lines (see Figs. 2.20 and 4.15): two are found in the spectrum of the I nuclei (^{1}H NMR) and two in the spectrum of the S nuclei (^{13}C NMR). The separation of the lines in the doublet is equal to the scalar spin-spin coupling constant J_{IS}; the centers of the doublets are the resonating frequencies (chemical shifts) f_I and f_S of the I and S nuclei, respectively. The spectrum results from transitions among the four possible energy levels of the spin system (Fig. 2.20). The four states (energy levels) of the IS system thus correspond to four possible combinations of states of the I and S spins. The first line in the spectrum of S (^{13}C) corresponds to the transition from the lowest energy level to the second energy level (transition $1 \rightarrow 2$), and the second line in the same spectrum corresponds to transition $3 \rightarrow 4$. Analogously, the lines in the spectrum of I correspond to transitions $1 \rightarrow 3$ and $2 \rightarrow 4$. The first S line originates in those molecules that have I nuclei (^{1}H) in the $< +\frac{1}{2} >$ state, and the second S line comes from molecules with spin I in the $< -\frac{1}{2} >$ state (orientation). In the same way, the lines of the I nuclei (^{1}H) spectrum come from molecules that have S (^{13}C) nuclei with spins in the $< +\frac{1}{2} >$ and $< -\frac{1}{2} >$ state, respectively.

Transitions among energy levels (or states) are caused by the rf pulse. We have seen that the pulse creates changes in population levels and creates coherence in the spin motion on the two energy levels connected by the transition. Accordingly, we can divide the magnetization of the spins M^S, into two components; the M^{S+} and M^{S-} magnetizations, which are vector sums of the nuclear magnetic moments of the S nuclei in molecules having their spins I in the $< +\frac{1}{2} >$ and $< -\frac{1}{2} >$ states, respectively. (For brevity, we sometimes designate the two spin-$\frac{1}{2}$ states by writing only $< + >$ and $< - >$ signs.) Analogously, the magnetization M^I can be broken into M^{I+} and M^{I-} components according to the individual states of the coupled S nuclei. In the pulsed experiments we rotate the magnetization from its equilibrium position along the z axis into the x, y plane and follow its rotation in this plane. The components of M^{S+} and M^{S-} magnetizations in the x, y plane correspond to coherence between the 1–2 and 3–4 states in a quantum statistical treatment of the experiment.

Let us see how our spin system behaves under the influence of various pulse sequence elements.

A.4.1.   Radio-frequency Pulses

A pulse is characterized by its radiofrequency, whether it is a pulse that affects the observed nuclei, or whether the rf field has the frequency of the other nuclei (I), which are usually irradiated through the decoupling channel. The two types of pulses are distinguished by specification of the nuclei affected (S or I). In linear schemes the nucleus is usually specified as an index in parentheses; for example, the symbology 90°(S) denotes a 90° pulse with a frequency close to the resonance frequency of the S nuclei. In graphical schemes the S or I label is appended to the row showing the events in one spectrometer channel (see, e.g., Fig. 2.5). The upper row in the graph often indicates the events taking place in the decoupler channel, while the lower row shows those in the observation channel. A pulse is represented in the graphical scheme by a rectangle. You should realize that the rectangle aptly describes the open and closed states of the gate in the rf circuitry. In reality, when the gate is open, the sample is exposed to rf irradiation. The rectangle is "filled" with oscillations; for example, when the frequency is 100 MHz, the magnetic field B_1 oscillates 1000 times during the pulse, which has a duration of only 10 μs.

The effect of a pulse on the spin system depends on the amplitude of the oscillating field, on the pulse length and phase, and on the state of the spin system at the beginning of the pulse. We have already discussed how amplitude B_1 and pulse length t are collectively described by the flip angle [Eq. (A.3-2)] as well as by the notation used to designate the pulse phase (see Section A.3). We should add, however, that we consider a pulse to be an x pulse if the pulse rotates the magnetization of nuclei with a positive magnetogyric ratio around the $-x'$ axis of the rotating frame of reference. For example, a $90°_x(S)$ pulse rotates the magnetization of the S nuclei from the direction of the $+z$ (or $+z'$) axis into the direction of the $+y'$ axis.

The required flip angle can be achieved by different combinations of rf field amplitude and pulse length. There are a number of important pulse experiments in which the flip angle does not provide sufficiently detailed information about the pulse length and rf field amplitude or power. The most common reason is that the pulse length determines the selectivity of the pulse. Selectivity means the spectral width that is affected by the pulse. Although there is only one transmitter frequency gated "on" during the pulse, the finite length of the pulse has the effect that spectral analysis (Fourier transform) of the pulse shows that there is a full range of frequencies present in the pulse. Roughly, the longer the pulse length (in seconds), the narrower is its spectrum, that is, the narrower is the range of frequencies around the transmitter frequency that are present in the pulse. This is also the range of frequencies of magnetization that are affected by the pulse. Hence a longer pulse is more selective than a shorter one. If we want to use a pulse to affect a broad spectrum of frequencies, we have to use a strong short 90° pulse that is not selective. To affect selectively only a limited spectral region by using a pulse with a specified flip angle, we must set the transmitter frequency close to the middle of the region and adjust the rf power (amplitude) so that the given flip angle is achieved with a pulse length that corresponds to the region to be affected. Several pulse sequences have been described for pulse calibrations, and these sequences allow determination of the flip angle (Section A.10). The spectrometer controls the rf power in units of decibels (dB), and it is important to note that the number of the decibels is $10 \log(P_1/P_2)$, where P_1/P_2 is the power ratio and not the magnetic field amplitude ratio. An increase in power of 6 dB means that the power has been multiplied by a factor of 4, but this increase of power corresponds to an increase of the B_1 amplitude by only a factor of 2.

The orientation of magnetization after the pulse depends not only on the parameters of the pulse that we have just discussed, but also on the initial orientation of the magnetization at the beginning of the pulse. For example, a $90^\circ_{x}(S)$ pulse (i.e., a pulse that rotates the magnetization of the S nuclei around the $-x'$ axis by 90°) does not turn the x component of magnetization, but it does turn the z'-axis and y'-axis components by 90° in the y',z' plane. We can obtain the effects of the other pulses by a cyclic permutation of the axes.

Thus far we have considered ideal pulses that create a homogeneous rf field over the whole sample and have the rf field amplitude and phase held constant for the entire duration of the pulse. However, real pulses produced in the spectrometer probe are not at all ideal. Moreover, even an ideal pulse would not produce perfectly the effects that we expect if its frequency is too far removed from the resonating frequency (off-resonance effects). Fortunately, Levitt and Freeman [265] have found that it is possible to design composite groups of several pulses in such a way that some single-pulses imperfections are compensated. With the shortest practical delay between pulses the group acts as a single composite pulse. The construction of a composite pulse depends on which imperfections have to be compensated and in what experiment. Various possibilities of constructing and using composite pulses were recently reviewed [222].

A.4.2. Delay without Decoupling

This pulse sequence element is characterized by its duration time. Its presence in a pulse sequence scheme does not require much explanation. In addition to preparation, evolution, and detection periods, which have a clear function in 2D NMR pulse sequences, there are usually delays whose function is to permit the magnetization components to achieve a certain phase relationship (e.g., to become parallel).

When there is no rf field acting on the sample, the magnetization precesses and relaxes unless it is in its equilibrium position. The projection of the magnetization $\bar{M}$ onto the x', y' plane ($M_{x,y}$) has a relaxation time T_2, and the z'-axis component (M_z) has a relaxation time T_1. In our discussion we shall not consider relaxations, since their effect on 2D spectra is usually not decisive (except in cases when 2D methods are used to measure relaxation times). In some cases, however, the relaxation places limits on the sensitivity, on the resolution that can be achieved, and on the usable length of the pulse sequence. (The spin system must not relax completely during the execution of the pulse sequence.) If we neglect relaxation, the M_z component, which is parallel to the magnetic field $\bar{B}$, does not change during the delay. The component in the x', y' plane, $M_{x,y}$, rotates, and the frequency of rotation depends on the spin system studied. In our spin system IS, the M^{S+} and M^{S-} components rotate with frequencies $f_S + J/2$ and $f_S - J/2$ with respect to our fixed laboratory frame of reference. (For clarity, we omit the index IS of the coupling constant J.) In the doubly rotating frame, which rotates with a frequency of f_S for spins S (we choose the frame of reference so that it rotates with this frequency), the two magnetization components rotate with frequencies of $+J/2$ and $-J/2$. If the two components were parallel at the beginning of the delay, they diverge or fan out during the delay. The corresponding magnetizations of the I nuclei, M^{1+} and M^{1-}, rotate with the same frequencies in the doubly rotating frame, since this frame rotates with the frequency f_I for nuclei I.

In many pulse sequences we need a delay in order to achieve a particular angle or phase relationship between the magnetization components of nuclei of the same type. For example, we need to reach an angle of 180° between the M^{1+} and M^{1-} components for polarizatison transfer (Section 2.4.2) in order to have a maximum efficiency of the transfer. This task is simple if we are concerned with the magnetization of spins I in the spin system IS, I_2S, I_3S, ..., I_nS. In such cases the M^1 magnetization always has only M^{1+} and M^{1-} components. (Remember that these components come from molecules in which the spins S are in $<+\frac{1}{2}>$ and $<-\frac{1}{2}>$ states.) The two components rotate in the x', y' plane with $+J/2$ and $-J/2$ frequencies (in hertz), and hence during a delay of t (seconds) the angle between them changes by $360 \cdot t \cdot J$

(degrees). This is why delays immediately preceding a polarization transfer pulse are the same for all common spin systems. The situation is more complex if we are concerned with the magnetization of S nuclei in such systems. Depending on the possible number of spin states of I nuclei, the magnetization M^S decomposes into $n + 1$ components (i.e., into the familiar doublets, triplets, and quartets observed in coupled spectra of spin systems $I_n S$ with $N = 1, 2,$ and 3). Each of these components rotates in the rotating frame of reference with a frequency $\Sigma m_I \cdot J$, where Σm_I is the sum of the magnetic numbers of the I nuclei that are present in the molecules responsible for the given magnetization component. Thus one of the inner more intense lines in a quartet is due to molecules with one spin of the I nuclei in a $< -\frac{1}{2} >$ state and the remaining two spins of the I nuclei in a $< +\frac{1}{2} >$ state. This line has a frequency of $+J/2$. The second of the inner lines has a frequency of $-J/2$, and the outer (weaker) lines of the quartet have frequencies of $+3J/2$ and $-3J/2$. Therefore, the required angle cannot be achieved simultaneously by all pairs of magnetization components, and thus we must find an optimum delay for the best overall performance of the experiment. These considerations are the basis of the dependence of the signal intensity on the length of the d_2 delay shown in Fig. 4.3.

A.4.3. Delay with Continuous Decoupling

A delay with continuous decoupling must be characterized not only by its length of time but also by the decoupling method (selective, coherent, broadband, etc.)

Decoupling of the I nuclei (i.e., irradiation of the sample with a frequency close to the f_I frequency) enhances the difference in the populations of the energy levels of S nuclei that are coupled with I nuclei and thus enhances the signal in the spectrum of the S nuclei by the nuclear Overhauser effect [266]. This, of course, pertains only to nuclei that have a positive magnetogyric ratio and $\gamma_I > \gamma_S$. This is the usual role of decoupling during the preparation period. The most important effect of decoupling for 2D NMR measurements is that during the decoupling of the I nuclei, the components of magnetization of the S nuclei, M^S, all rotate with the same frequency, f_S. That is, they are stationary in the rotating frame. In our example of chloroform, the two M^{S+} and M^{S-} components describe the same angle at the end of the delay with decoupling as they had encompassed when the decoupling was switched on, which is, of course, the basis of the frequently used spectral simplification by decoupling during the detection period. Since the magnetization components rotate with the same frequency, only one line is found in the spectrum for one type of S nuclei.

A.4.4. Magnetic Field Gradient

The application of magnetic field gradients is an essential feature of the types of 2D (and 3D) NMR spectroscopy known as imaging, but pulses of magnetic field gradient in the z-axis direction are also employed in the pulse sequences that are of concern here. The gradient pulses usually serve to remove certain undesirable magnetization components (see 2D exchange spectroscopy, Section 4.3.1). The unwanted transverse components of magnetization are removed as the spins loose their coherence in the inhomogenous magnetic field. The coherence can be regained if the same gradient pulse is applied to the spins, either with inverted polarity or with the same polarity, but after the direction of rotation of the magnetization components has been inverted. Magnetic field gradient pulses can thus replace complicated phase cycling schemes in selecting specific coherence-transfer pathway (Section 4.2.1 and A.7, e.g., in N-type detection in COSY spectra). The use of gradient pulses for this purpose in ^1H NMR leads to considerable savings in instrument time, since the lengthy time averaging needed to complete the phase cycle for protons is not necessary for improvement of the signal-to-noise ratio [267].

The simplest way to produce a magnetic field gradient is to have a direct current pulse through the homogeneity "shimming" coils. More complicated approaches include probe inserts with specially designed gradient coils. Since the gradients obtained by these methods might not be sufficient, and since the gradient pulses in any case will disturb the frequency-field lock, alternative means for coherence-transfer pathway selection are being sought [268]. There is no established way of denoting the use of a magnetic gradient in a pulse sequence scheme.

A.4.5. Detection

Some authors denote detection (acquisition) by a symbolic FID in graphical schemes; we simply use a triangle. In some experiments the receiver phase (or its cycling) must be specified, for example, for selecting a particular type of detection (such as quadrature detection), as discussed in Sections A.2 and A.7. Obviously, signal detection in the observation channel can be accompanied by sample irradiation through the decoupler channel. In such a case the decoupling method should be specified, just as it is in the case of a delay with decoupling. For special experiments that use delayed decoupling (the decoupling begins after some part of the FID has been detected without decoupling [269]), the delay must also be specified.

A.5. ELEMENTARY PULSE SEQUENCES AND PULSE CLUSTERS

We have already mentioned that all pulse sequences can be constructed from the few pulse sequence elements listed in the preceding section, but it is obvious that the useful pulse sequences contain characteristic clusters of elements. These clusters perform some rather complex tasks, and we shall examine in some detail four such clusters: spin-echo, INEPT, DEPT, and BIRD.

A.5.1. Spin-Echo

We shall start our discussion of clusters of pulse sequence elements with spin-echo. Spin-echo is one of the most usefuls tools of pulsed NMR spectroscopy. It was discovered, explained, and put to good use many years ago by Hahn [270]. Although it can be used as a pulse sequence of its own, spin-echo is incorporated into many 2D pulse sequences in order to serve certain purposes that will become clear later.

We call a signal an "echo" when the maximum signal does not occur immediately subsequent to the pulse, but instead at some later time. A spin-echo appears if some pulses are repeated; various combinations of pulses lead to an echo. We shall discuss only the simplest type of a spin-echo pulse sequence, one that is easy to understand and gives easily predictable results. This useful pulse sequence can be described by the linear scheme

$$90^\circ_x(S) - t_D - 180^\circ(S) - t_R - \text{detection}(S) \qquad (A.5\text{-}1)$$

Note that we have left the phase of the 180° pulse unspecified at this point. The first pulse turns the magnetization of the S nuclei into the direction of the y' axis of the rotating frame of reference. (We assume that the magnetization was at its equilibrium position before the pulse.) During the delay (without decoupling) the components of magnetization M^S diverge. For simplicity, we shall consider only two components of M^S. The one that rotates faster will be labeled f, and the other one, the slower one, will be called s.

The two components (see Fig. A.7) can be our familiar two M^{S+} and M^{S-} components of magnetization M^S in the IS spin system, or they can be magnetizations M^S from two different types of S nuclei (with a chemical shift difference), or they can be magnetizations M^S of the one type of S nuclei but from different sample regions. (Thus they would be subject to a different magnetic field B_0 and hence have different precession frequencies. The different frequencies can be either artificially introduced if a magnetic field gradient is applied to the sample during the delay, or they can be caused simply by a magnetic field inhomogeneity.)

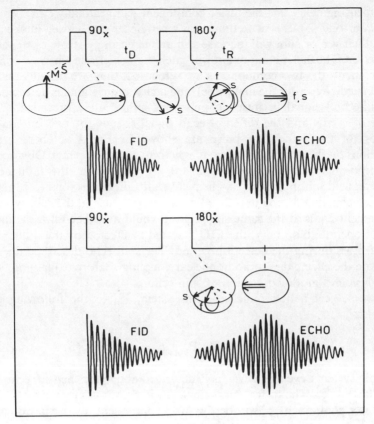

Figure A.7 Spin-echo: pulse sequences, fast (f) and slow (s) magnetization components, and observed signals for two different phases of the refocusing pulse.

The two components, which are initially parallel, will diverge or fan out during the first delay t_D. At the end of this defocusing delay, t_D, the two components encompass a certain angle.

The second pulse turns both components by 180°; the exact orientation of the two components depends on the phase of this second (refocusing) pulse. In any case, however, immediately after the refocusing pulse, the faster component f will now lag behind the slower one. The component that we labeled "faster" (f) will now be as much behind the "slower" one s as the slower one was behind the faster one before the refocusing pulse. (Note that the angle between the two components has not changed by the second pulse; the only change is that the slower component is now ahead of the faster component in the rotating frame of reference.) During the second delay (i.e., the refocusing delay, t_R) the faster component will gradually catch up with the slower one. As the two components get closer and closer in the x', y' plane, their vector sum will increase and an increasingly stronger signal will be detected. The faster component will catch the slower one when $t_R = t_D$. At that moment, the two components are refocused, they are parallel (as they were immediately after the first pulse), and a maximum signal is detected. It is a "spin-echo" signal because no pulse has immediately preceded it. The course of events and the different echo signal shapes for two orientations (phases) of the refocusing pulse are shown in Fig. 4.7. There are no limitations on the origin of the two magnetization components. Obviously, a spin echo-pulse sequence will éliminate, by refocusing, the influence of magnetic field inhomogeneities, chemical shift differences, or heteronuclear spin-spin coupling during time $2t_D$. Because of refocusing, the spin system at the moment $2t_D$ is at the same state that it would have been if these factors did not operate. Naturally, the FID detected is affected by the influence of these factors during the detection period. The spin-echo described does not eliminate the effects due to homonuclear coupling. Before explaining, let us first consider a modified version of spin-echo.

If we subject a heteronuclear spin system, IS, to the following pulse sequences:

$$90^\circ_x(S) - t_D - 180^\circ(I) - t_R - \text{detection(S)} \qquad (A.5\text{-}2)$$

The spin system develops as we have described above for the spin echo until the 180°(I) pulse. The faster component, f, will be our component M^{S+}, and the slower component, s, will be the M^{S-} component, as discussed earlier. The 180° pulse will not change the position of the two components since it does not act on the S nuclei. It will, however, turn the orientation of the I nuclei, with which our observed S nuclei are coupled. Those I nuclei which are in the $< +\frac{1}{2} >$ state before the pulse will now be in the $< -\frac{1}{2} >$ state after the pulse, and vice versa. This relabeling of the I spins will also affect our M^S components. The 180°(I) pulse changes the faster component M^{S+}, into the slower one, M^{S-}, and similarly, the M^{S-} component is changed into the faster component. Of course, the positions of the components immediately after the second pulse are not altered; the components are merely relabeled. However, this also causes an echo formation because the component that was behind is changed into the faster component and catches up with the slower one (which was the more advanced immediately after the second pulse). It is clear that this modification of spin-echo eliminates or refocuses the influence of heteronuclear spin coupling, but it does not refocus chemical shift or magnetic field inhomogeneity effects.

In the case of pulse sequence (A.5-1) acting on a homonuclear spin system

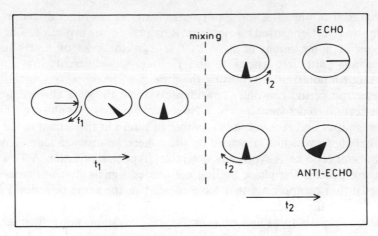

Figure A.8 Formation of a coherence transfer echo.

SS′ with homonuclear coupling, the refocusing 180° pulse acts on all spins in the system. It turns magnetization components by 180° in the x', y' plane (and thus causes a refocusing of the effects of chemical shifts and field inhomogeneity). It also brings about a relabeling of all spins just as in the modification of spin-echo above. The outcome of the two effects described for a 180° pulse is that the faster component, after the pulse, is behind the slower one, but at the same time, it is changed into the slower component. Similarly, the slower pulse component is placed ahead of the faster one and is changed into the faster one. Although these complicated changes have taken place, the magnetization components remain effectively in their same positions in which they were before the "refocusing" pulse. Thus the effects of homonuclear coupling are not eliminated by refocusing.

An interesting and important type of echo, the coherence transfer echo, is formed in practically all correlated 2D NMR experiments. During the detection period of a 2D NMR experiment, we record a signal that has a frequency f_2. Part of this signal comes from the magnetization that has evolved during the evolution time with a frequency f_1. This signal leads to a f_1, f_2 cross-peak in the 2D spectrum. Because of magnetic field inhomogeneity, the magnetizations fan out during any delay, as suggested in Fig. A.8 by the presence of shadowy areas.

When the direction of rotation of the magnetization with frequency f_1 during the evolution period is opposite to that of the magnetization with frequency f_2 during the detection period, an echo is also formed. The magnetic field inhomogeneities that fanned out the magnetization during evolution will cause their refocusing during detection. The component that rotates faster in one direction also rotates faster in the other direction, and thus in the detection period it loses what it gained during evolution. The echo signal is at a maximum at $t_R = t_1 \cdot f_1 / f_2$. In the case of heteronuclear correlations (with frequencies f_1 and f_2 markedly different) that involve coherence (or polarization) transfer, the time t_R of the echo maximum during the detection period can differ considerably from the time allowed for the magnetization to defocus ($t_1 = t_D$). By a suitable choice of phase cycling of the transmitter and receiver, it is possible to retain in the accumulated FID only that signal which is refocused by the coherence transfer echo. Thus we speak about echo or N-type peak detection (Appendix Section A.7). In the averaging process the phase cycling eliminates signals that are caused by magnetization components that have rotated in the same direction during both the evolution and the detection period. Signals from these components, which are referred to as anti-echo or P-type peaks, are not refocused, and thus the signal-to-noise ratio of these signals is somewhat degraded by the effects of magnetic field inhomogeneities.

A.5.2.   INEPT and DEPT

The pulse sequences known as INEPT (Insensitive Nuclei Enhanced by Polarization Transfer) and DEPT (Distortionless Enhancement by Polarization Transfer) dramatically enhance the signals of nuclei with low magnetogyric ratios. The polarization transfer experiments described earlier (selective polarization transfer and heteronuclear chemical shift correlations, Sections 2.4.1 and 4.1.1, respectively) are sensitive to proton chemical shifts; the extent of polarization transfer depends on the proton chemical shift. The enhancement of the INEPT and DEPT signals, however, depends not on the proton chemical shift but rather on the coupling constant and multiplicity of the signal of the observed nucleus. Therefore, INEPT and DEPT are two pulse sequences that are often used in 1D NMR spectroscopy for signal enhancement of less receptive nuclei and for editing spectra (especially ^{13}C NMR spectra) according to the multiplicity of the lines. These uses of INEPT and DEPT have been reviewed on several occasions [3, 271, 272].

In 2D NMR spectroscopy the INEPT or the DEPT pulse sequence usually replaces the single pulse at the end of preparation period in order to increase the sensitivity of methods with an inherently low sensitivity (e.g., correlations via small coupling, multiquantum filtered experiments, etc.).

The older of the two sequences, INEPT, has undergone several changes [55–57] since its invention by Morris and Freeman in 1979 [54]. Its most complete form can be summarized by the following pulse scheme [271]:

(I)	$90^\circ_x - \tau - 180^\circ - \tau - 90^\circ_y$		$-d_2/2 - 180^\circ - d_2/2$	$-$ (decouple)
(S)	180°	90°	180°	$-$ detect
	polarization transfer *period*		*refocusing* *period*	*detection period* *with/without decoupling*

$$(A.5\text{-}3)$$

where the total polarization transfer time should be $2\tau = 1/(2 \cdot J_{IS})$, and the refocusing delay d_2 chosen according to the value of the coupling constant J_{IS} and signal multiplicity, just as the corresponding d_2 delay was chosen for heteronuclear correlations (see Section 4.1.1). For a spin system of the type $I_n S$, the optimum delay d_2 is [58]

$$(d_2)_{opt} = (180^\circ \cdot J_{IS})^{-1} \arcsin(n^{-1/2}) \qquad (A.5\text{-}4)$$

The detailed explanation of INEPT in terms of the vector model is given in the original paper [54]; it is not necessary to repeat it here. Since the steps of which the INEPT sequence is composed have already been considered, it should suffice that we recall some aspects that we have discussed in the section on the fundamentals of heteronuclear chemical shift correlations (Section 2.4.2) and in the section on decoupling in the detection period of these experiments (Section 4.1.1).

The polarization transfer period in INEPT has essentially the same function as the evolution period in those heteronuclear correlation experiments cited above, except that a pair of refocusing 180° pulses has been introduced into the middle of the period. The refocusing pulses ensure that the effects of chemical shifts (of I spins) are eliminated by refocusing, but at the same time the M^{1+} and M^{1-} magnetization components due to J_{IS} coupling continue to diverge for the rest of the period. After a properly chosen polarization transfer time of 2τ, the two components are aligned in opposite directions along the x axis. For any type of I nuclei in the sample, the two magnetization components are in the optimum position for maximum polarization transfer, which is accomplished by the pair of 90° pulses. Naturally, the enhancement of the signal of S nuclei depends on the number of I nuclei coupled to S (for exact formulas, see references [58, 274]). The signal enhancement is well apparent from comparison of Fig. A.9a and b; for the same signal-to-noise ratio it is sufficient to average only one-fourth of the scans needed for the normal spectrum. (The delays and flip angles were not selected for spectral editing; hence all protonated carbon lines are visible in the figure.) The refocusing period has exactly the same role as the d_2

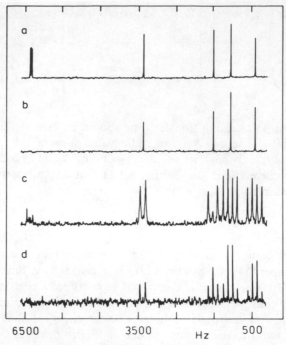

Figure A.9 ^{13}C spectra of 2-butanol (10%) in hexadeuteriobenzene: (*a*) time average of 32 transients measured by pulse sequence of Fig. 2.1 with proton decoupling; (*b*) time average of 8 transients measured by INEPT pulse sequence (A.5-3) with proton decoupling using delays τ = 1.92 ms and d_2 = 1.92 ms; (*c*) time average of 32 transients measured by INEPT pulse sequence (A.5-3) without proton decoupling but with refocusing, the same delays as in (*b*), absolute-value presentation; (*d*) same conditions as in (*c*) except that the DEPT pulse sequence (A.5-5) was used with θ = 45°, phase-sensitive presentation.

delay in the mixing period of heteronuclear chemical shift correlations (Section 4.1.1); the dependences shown in Fig. 4.3 are used for selection of the proper value of d_2. If spectra are to be measured without decoupling, the refocusing period can be deleted; however, some lines will then show inverted intensities, which is again similar to the case of the analogous heteronuclear correlations. The lines within multiplets of S nuclei will have relative intensities different from those obtained in a normal acquisition (Fig. A.9*c*).

Such intensity distortions are not observed when the DEPT sequence discovered by Doddrell et al. [60] is employed (Fig. A.9*d*). Moreover, the DEPT pulse sequence contains fewer pulses and is therefore less sensitive to pulse imperfections and less demanding on pulse programming. The DEPT sequence is

(I) $90^\circ_x - 2\tau - 180^\circ_x - 2\tau - \theta^\circ_y - 2\tau -$ | $-$ (decouple)

(S) 90° 180° | $-$ detect

polarization transfer | *detection period*
period | *with/without decoupling*

(A.5-5)

with the delays τ chosen as above for INEPT, and the flip angle θ° of the last (I) pulse playing the same editing role in DEPT as the d_2 delay does in INEPT (for nuclei with spin $\frac{1}{2}$) [273]. The flip angle θ° should be chosen so that

$$\theta^\circ = 180^\circ \cdot J_{\mathrm{IS}} \cdot d_2 \qquad (A.5\text{-}6)$$

where d_2 is the optimum d_2 delay for the INEPT experiment.

Although attempts have been made to provide a description of the action of the DEPT pulse sequence in terms of the vector model [59, 107, 289], an adequate description is beyond the scope of this book. We must merely accept that DEPT, at the beginning of the detection period, provides aligned magnetization components from all S spins and with proper intensity ratios (Fig. A.9d).

When the two sequences are utilized for increasing the sensitivity of a 2D NMR method, only their polarization transfer portions are inserted into the 2D pulse sequence. When needed, however, the refocusing part can also be incorporated in order to prepare aligned magnetization components. It should be noted that those applications used to increase the selectivity of measurements will require a more careful evaluation for proper incorporation into the pulse sequence (see Sections 3.1.3 and 4.1.1 for references).

A.5.3.   BIRD

In contrast to spin-echo, which has been known for many years, the bilinear rotation decoupling (BIRD) pulse cluster, or Pines operator, was devised rather recently [275]. The BIRD pulse cluster has a rather narrow range of applications, but it has rapidly found its way into many 2D NMR pulse sequences. It serves the following purposes: (1) it provides proton homonuclear broadband decoupling, (2) it eliminates heteronuclear couplings, and (3) it discriminates between direct and long-range heteronuclear couplings.

The simplest BIRD cluster of pulses that performs these tasks is

$$90^\circ_x(^1H) - d - 180^\circ_x(^1H), 180^\circ(^{13}C) - d - 90^\circ_{-x}(^1H) \quad \text{(A.5-7)}$$

where $d = 1/(2 \cdot {}^1J_{CH})$. Also, we specify ^1H and ^{13}C in the BIRD sequence (i.e., $I = {}^1H$ and $S = {}^{13}C$) since BIRD is not likely to be applied to other nuclei.

We shall illustrate the use of the simple BIRD cluster given above on a typical spin system; a few obvious modifications will be mentioned in passing. Those who are interested in the more general forms of the Pines operator (e.g., with different flip angles of the first and last pulses [266] or in pulse clusters that are compensated for the distribution of coupling constant values [275–277]) should consult the references listed.

The action of a BIRD pulse cluster is easy to understand since it is essentially a spin-echo pulse sequence (1) to which one pulse is added at the time of maximum echo refocusing, and (2) which has specifically chosen lengths of defocusing and refocusing delays. Recently, a theoretical analysis was published by Rutar and Wang [278].

Let us consider first the effect that a BIRD pulse cluster has on (remote) protons that are not directly coupled to ^{13}C carbons. If we recall that long-range C–H couplings are an order-of-magnitude smaller than one-bond couplings, we realize that the magnetization components due to long-range coupling do not appreciably diverge from each other during the short delays noted as d in the BIRD pulse cluster. Therefore, we can ignore the action of the $90^\circ(^{13}C)$ pulse on such protons. According to our discussion of spin-echo (see the bottom part of Fig. A.7), proton chemical shifts are refocused along the $-y$ axis just before the last pulse of the BIRD cluster. The last pulse turns the magnetization of these protons in the direction of the $-z$ axis; that is, the BIRD cluster has inverted the spins of the protons that are remote from carbon-13.

Close protons, those directly coupled to the ^{13}C nuclei, are affected differently by the BIRD pulse cluster. The $180^{\circ x}(^1H)$ pulse in the middle of the BIRD cluster turns the two magnetization components due to direct coupling, M^{1+} and M^{1-}, by 180° in the x', y' plane. Thus the defocusing effects of chemical shifts and inhomogeneities on the magnetization during the first delay d are overcome at the end of the second delay d by a refocusing along the $-y$ axis as in the case above. The simultaneous $180^\circ(^{13}C)$ pulse relabels the two components as in the case of spin-echo acting on a homonuclear spin system and thus inverts their directions of rotation. The delays d were chosen so that the last pulse finds the components refocused along the $+y$ axis instead of along the $-y$ axis (i.e., because of relabeling, each of the components has diverged from the refocusing $-y$ direction by 180°). Finally, the last pulse rotates the magnetizations back into their original starting positions (i.e., the BIRD cluster does not affect a proton bound to carbon-13).

We have thus far tacitly assumed that the magnetizations were oriented along the z axis prior to the beginning of the BIRd pulse cluster, although we could maintain the same elements of discussion with any starting orientation. In the final analysis, the BIRD cluster inverts the spins of remote protons and leaves unchanged the spins of close protons. For this reason BIRD can be easily incorporated into any pulse sequence. It will replace an ordinary 180° pulse with a pulse that has the three effects noted in the first paragraph of this section.

Again, when BIRD is used to replace an ordinary refocusing 180° pulse in the middle of some 2D pulse sequence, it leaves unaffected the spins of close protons, but it inverts the spins of remote protons and of carbon-13 nuclei. Consequently, the components of magnetization of close protons, which had fanned out during the first half of the evolution period because of heteronuclear and homonuclear couplings, are refocused by the BIRD pulse at the end of the evolution period. Therefore, the resulting 2D spectra are proton-proton and proton-carbon decoupled in the direction of the f_1 axis, but they show the chemical shifts of the protons. (Naturally, when two nonequivalent protons are on the same carbon-13, such as the diastereotopic protons of the CH_2 group in 2-butanol, they are not affected by the BIRD cluster and the coupling, usually strong, is retained; see Fig. 4.6.)

Obviously, the effect of the BIRD pulse cluster can be reversed by inversion of the relative phase of one of the pulses (e.g., if the last pulse is a x pulse instead of a $-x$ pulse, the modified BIRD cluster does not affect the distant protons but inverts the spins of close protons, etc.).

BIRD clusters have been employed in heteronuclear chemical shift correlations utilizing short-range and long-range correlations (Section 4.1.1 [66, 89–95]), heteronuclear J resolved spectra (Section 3.1.1 [116, 275]), and others. Problems connected with the use of imperfect pulses [279] and the effects of strong coupling [280] have been considered in some detail in the literature.

A.6. FUNDAMENTAL 1D EXPERIMENTAL CONSIDERATIONS

To have a better understanding of some of the experimental problems encountered in 2D spectroscopy, we shall briefly review a few related aspects of 1D measurements. These and many other aspects of pulsed NMR experiments are considered in detail in several monographs [4, 5, 281, 282].

We usually strive for well-resolved spectra with a good signal-to-noise ratio, the proper representation of line frequencies, and the shortest possible measurement time. To satisfy these demands, the experiment should be "time-optimized," and the digital aspects of pulsed Fourier transform NMR, which we have thus far ignored, must also be considered.

To have spectra with absorption linewidths (LW), the detection (or acquisition) time t_2 must be

$$t_2 \geq 1/LW \qquad\qquad (A.6\text{-}1)$$

For example, to resolve two lines 0.5 Hz apart the FID data must be recorded for a detection time of at least 2 seconds.

The line frequencies will be properly represented in a spectrum (that is, the spectrum appears without fold-over or image reflection lines) when the transmitter (carrier) frequency is placed either to one side of the spectrum (when single channel detection is used) or into the middle of the spectrum (when quadrature detection is used). It is also necessary that the total spectral width (SW) be correctly estimated. An underestimation of the spectral width leads to a "folding in" of the lines that are outside the estimated spectral width. When single channel detection is used, a line that should occur in the spectrum at a frequency of $(SW + f)$ Hz (that is, f Hz outside the estimated spectral width), is actually found in the measured spectrum at a frequency of $(SW - f)$ as if it were "folded" back around the edge of the spectrum. An overestimation of the spectral width results in an inefficient use of memory and a deterioration of the signal-to-noise ratio.

The estimated spectral width is used for the calculation of the rate at which data should be sampled. This calculation requires some explanation. The FID shown in Figure A.3 appears as a smooth curve, but it is stored in the computer as a series of discrete values, and the smooth curve appearance is the result of a clever interpolation between discrete points. The discrete nature of the data becomes obvious if the time scale of the FID is expanded as shown in Figure A.10.

The continuous (analog) FID signal from the receiver is sampled with a predetermined frequency at equidistant points and converted into digital form by an analog-to-digital (ADC) converter. The digitalized values of the signal are stored in the computer memory for further processing. Proper spectral representation of frequencies requires the use of a sampling frequency

$$f_{SA}^{SD} \geqslant 2 \cdot SW \qquad\qquad (A.6\text{-}2a)$$

for single-channel detection and

$$f_{SA}^{QD} \geqslant SW \qquad\qquad (A.6\text{-}2b)$$

for quadrature detection. It would be useful to express these conditions in

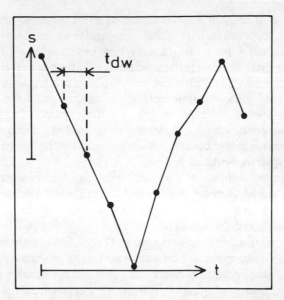

Figure A.10 First portion of the FID from Fig. A.3: 20-fold expanded time scale. The time interval between two consecutive points is the dwell time, t_{dw}, which is the reciprocal of the sampling frequency, f_{SA}.

terms of dwell time, t_{dw}, which is the interval between the data pairs or the reciprocal of the sampling frequency. The conditions for proper frequency representation are then

$$t_{dw}^{SL} \leq 1/(2 \cdot SW) \qquad (A.6\text{-}3a)$$

$$t_{dw}^{QD} \leq 1/SW \qquad (A.6\text{-}3b)$$

Hence, when the data are acquired for a period of $t_2 = 1/LW$ with the given sampling frequencies, we obtain $N = (2 \cdot SW/LW)$ data points with single channel detection and $N = SW/LW$ complex data points with quadrature detection. (Since the complex data points require two memory locations each, the two detection schemes have identical memory requirements for the same spectral width and spectral resolution in 1D spectroscopy.) The advantages of using quadrature detection in 1D experiments consist of a better use of transmitter power and a better signal-to-noise ratio.

We should note that we have neglected relaxation (T_2), which might place limits on the resolution in the above discussion of data requirements.

Consideration of relaxation rates is decisive in optimizing the time performance of the experiments. In simple one-pulse experiments the strongest signal is obtained after a 90° pulse. When time averaging is necessary to improve the signal-to-noise ratio, a better S/N ratio per unit time is obtained for a smaller flip angle. With a reduced flip angle, a higher pulse repetition rate can be used since the time needed for relaxation of the spin system is shorter. The signals detected in the repetitive passes through the pulse sequence (transients, scans) are constructively added in accordance with the pulse and receiver phase cycling, which also depend on the detection system used (see Appendix Sections A.2 and A.4.5).

The maximum number of transients that can be averaged for improvement of the signal-to-noise ratio is not unlimited; it is determined by the computer word length (bits), the digital resolution of the analog-to-digital converter (bits), and the signal intensity.

Before subjecting the time-averaged FID to a Fourier transform, the FID can be treated mathematically for additional signal-to-noise improvement or for increased resolution. The spectrum obtained by Fourier transform of a FID that had been multipled by a decreasing exponential of the form

$$\exp(-|at_2|)$$

has a better signal-to-noise ratio but the lines are broadened. In contrast, multiplying the FID by an ever-increasing exponential

$$\exp(+|at_2|)$$

leads to a spectrum with narrower lines but with a less favorable signal-to-noise ratio. A number of other mathematical functions have been proposed for FID multiplication in order to provide particular improvement in the final spectra. Resolution can also be improved by "zero-filling," which consists of adding zeros to the end of the FID (thus expanding the FID data twofold or more) prior to the Fourier transform. Various apodization or weighting functions, digital functions, and other mathematical means have been reviewed recently [260]. In every experiment a compromise between sensitivity and linewidth must be made, and possible lineshape distortion must also be considered. The optimum choice of weighting function depends on the purpose of the experiment and upon the repertoire of the functions available in the spectrometer software package. Usually, the best solution is found by trial and error, especially with an array processor for the fast Fourier transform routine.

A.7. PHASE CYCLING IN 2D EXPERIMENTS, QUADRATURE DETECTION DURING EVOLUTION, N- AND P-TYPE DETECTION, AND PURE ABSORPTION LINESHAPES

Although pulse and receiver phase cycling is an important and integral part of a pulse sequence, we have not yet given cycling "recipes." We have delayed the discussion of these topics to this point of the appendix for several reasons:

1. The correct explanation requires that we treat the signal as a complex exponential [1, 99] or use the density matrix [16, 186, 187].
2. Since phase cycling is designed to improve some of the practical aspects of 2D NMR experiments, it is not essential to understand phase cycling in order to grasp the fundamentals of 2D spectroscopy or to apply the principles of 2D NMR.
3. The reader is not expected to design his own pulse sequence but instead to use sequences provided as a part of a spectrometer software package (with phase cycling included).
4. Tables of phase cycling tend to be sizable as the number of pulses in the sequence increases.

In each chapter we have provided references that give phase cycling schemes. The phase cycling of your pulse sequence can be compared with those given in the literature by using the rules described in Appendix Section A.8.

In this section we offer an oversimplified explanation of the basic idea behind quadrature detection during evolution by phase cycling. We shall use a spin system consisting of isolated I nuclei only, and we shall choose the Jeener (H,H-COSY) two-pulse sequence for the model 2D experiment. Extensions to other spin systems and pulse sequences are left to the reader.

In our discussion of the Jeener experiment we assume quadrature detection during t_2; CYCLOPS phase cycling for artifact suppresion is mentioned later. The first 90°_x pulse turns magnetization M^1 into x', y' plane of the rotating frame of reference (which rotates with frequency f_r in the laboratory frame). During evolution time t_1 the magnetization travels an angle of $2\pi(f - f_r)t_1$. The purpose of introducing quadrature detection is to distinguish whether this angle is positive or negative, or whether the Larmor frequency f of I nuclei is higher or lower than the frequency of the transmitter, f_r (Fig. A.11). Only when we can distinguish higher and lower frequencies can we place the transmitter frequency in the middle of the spectrum (with the benefits already described in Sections 2.5, 4.1.1, and 4.1.2).

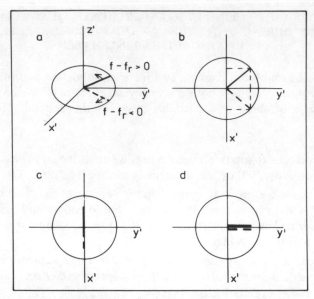

Figure A.11 Magnetization components rotating in positive and negative directions (with positive and negative frequencies): (*a*) the two magnetization components; (*b*) the two components in the x', y' plane; (*c*) the magnetizations left in the x', y' plane after a $90°_x$ pulse at $t_2 = 0$; (*d*) the magnetizations left in the x', y' plane after a $90°_y$ pulse at $t_2 = 0$. Solid line, magnetization with a positive frequency; dashed line, magnetization with a negative frequency.

The second 90° pulse applied at the end of the evolution period does not change the magnetization component along the direction of the pulse; it does turn the component perpendicular to the field of the pulse in the direction of the $\pm z$ axis. The pulse acts as a filter or detector and transmits information about the magnetization component in the direction of one axis only.

Thus if the second pulse is applied along the x' axis, it does not change the M_x magnetization component, and it turns the M_y component in the direction of the $\pm z'$ axis. (This component is responsible for the polarization transfer leading to cross-peaks; Section 2.4.2.) In our experiment with isolated I nuclei we are concerned with the M_x component at $t_2 = 0$, since this magnetization is left in the x', y' plane to rotate and to be detected during detection time t_2. Its magnitude is $M \sin 2\pi(f - f_r)t_1$ (see Fig. A.11). Both the intensity of the detected FID and also the intensity of the line obtained after the first Fourier transform will be proportional to this factor. Hence, after the transposition the interferogram intensity will also vary with the evolution time, that is, as $\sin 2\pi(f - f_r)t_1$ (amplitude modulation). As we have seen in discussion of quadrature detection during t_2 (Section A.2), a

Fourier transform of this dependence cannot distinguish positive and negative rotations. (This might appear surprising since the magnetizations shown in Fig. A.11 rotate in opposite directions and produce M_x components of opposite signs. One should realize, however, that M_x components of opposite signs can be also created at time t_1 by magnetizations rotating in the same direction but having an initial phase difference of 180°.)

This situation is analogous to single-channel detection during t_2. We have seen that the two directions of rotation can be distinguished if the signals detected along two perpendicular directions are taken together for one complex signal and the complex signal is then subjected to Fourier transform. (Note, however, that for the second Fourier transform a complex signal is always provided if a complex transform is used for the first Fourier transform. When properly phased, the real and imaginary parts of this complex signal correspond to absorption and dispersion modes along the f_2 axis.)

In the case of quadrature detection during t_1 the two perpendicular magnetization components at the end of the evolution time can be selected by using the above-mentioned properties of the second pulse. Quadrature detection during t_1 is accomplished by performing two experiments. In addition to the first experiment with the second 90° pulse in the direction of the x' axis, 90_x°, the FID is also registered in a second experiment using the mixing pulse in the y'-axis direction, 90_y°. The data obtained in the two experiments are then combined as discussed below.

Following the description above of the first experiment with the 90_x° second pulse, we find that the interferograms obtained in the second experiment with the 90_y° pulse vary with time t_1 as the M_y magnetization component, that is, as $M \cos 2\pi (f - f_r)t_1$.

When we combine the amplitudes of the FIDs (or line intensities) measured in the two experiments for the same time t_1, the phase of the magnetization at t_1 is fully determined (the two components M_x and M_y determine vector $M_{x,y}$ in the x', y' plane), and hence the direction of rotation is obtained after the second Fourier transform. There are, however, several ways in which the results of the two experiments can be combined to provide the complex signal for the second FT.

When considering some of the possible combinations of the signals obtained in the two experiments, we shall pay attention not only to the ability of the combination to distinguish frequency signs but also to the resulting lineshape. The most straightforward combination would be to store the FIDs of the two experiments separately, to process them separately, and finally to sum the two resulting 2D spectra. The final stage of such a processing is simulated in Fig. A.12 for the case of $f > f_r$ (positive rotation).

The 2D spectrum (imaginary part) obtained from the first experiment

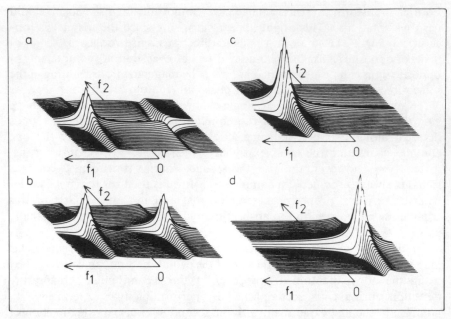

Figure A.12 Simulated 2D spectra obtained in the two experiments described in the text: (*a*) imaginary part of the spectrum obtained in the first experiment employing a $90°_x$ mixing pulse; (*b*) real part of the spectrum obtained in the second experiment employing a $90°_y$ pulse; (*c*) the result of addition of spectra *a* and *b*, P-peak; (*d*) the result of subtracting *a* from spectrum *b*, N-peak).

exhibits peaks at both $+(f - f_r)$ and $-(f - f_r)$ frequencies; both peaks have phase-twisted lineshapes. In agreement with sine modulation, the two peaks have opposite intensities (see Fig. A.4 for the FT of the sine function). The 2D spectrum (real part) produced by the second experiment also has two peaks with phase-twisted lineshapes, but the two peaks have intensities of the same sign (in agreement with the cosine modulation; see Fig. A.4). Obviously, neither of the two experiments alone discriminates the frequency signs, but we can take advantange of opposite peak intensities in the first spectrum and add the two spectra (i.e., we add the intensities at the corresponding points in the two spectra). We then obtain a spectrum containing only one peak and thus frequency sign discrimination is achieved. Since addition leads to a peak that has its frequency during time t_1 with the same sign that it has during the detection period (positive frequency, shown in Fig. A.12), those peaks detected in this manner are referred to as P-peaks and anti-echo peaks. Analogous subtraction of the two spectra also produces a 2D spectum with only one peak, but the peak appears to have a frequency with the opposite

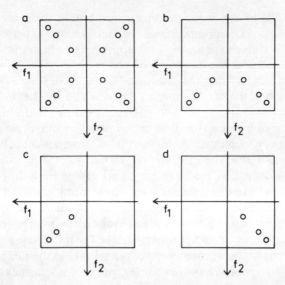

Figure A.13 Schematic 2D spectra showing the effect of quadrature detection: (*a*) single-channel detection in both axes, frequency signs cannot be determined; (*b*) quadrature detection during t_2, all three diagonal peaks have positive frequencies; (*c*) quadrature detection during the t_2 and P-type detection during t_1; (*d*) quadrature detection during t_2 and N-type detection during t_1.

sign (negative in our example) during the evolution period compared to the detection period. Such peaks are called N-peaks or echo peaks and are narrower than P-peaks if measured in an inhomogeneous magnetic field. A mathematical inversion of the direction of rotation has the same effect as the 180° pulse in a spin-echo pulse and leads to refocusing of the effects of magnetic field inhomogeneity. N- and P-peaks also have a different flip-angle dependence. The classification of N- and P-peaks is not limited to the homonuclear cases and to the single quantum correlations discussed here for illustration; the classification is readily extended to heteronuclear correlations and to multiple quantum spectra.

Relationships among the various types of detection are demonstrated by the schematic spectra (Fig. A.13) measured with single-channel detection and in quadrature in both dimensions using N- or P-type detection. Only three diagonal peaks are shown for simplicity; the transmitter (carrier) frequency is placed at the low-frequency end of the spectrum.

The data processing required is cumbersome, requires a large computer memory, and yields spectra with phase-twisted lineshape. The first two obstacles can easily be overcome by utilizing the linearity of Fourier transform. Instead of combining the two 2D spectra themselves, we can combine the FIDs of the two experiments and obtain the same result (thus saving data space and computing time). In fact, the two experiments can be combined into a single experiment with appropriate phase cycling.

The difficult point is the designing of phase cycling that would combine FID data in such a way that the resulting 2D spectrum would be equivalent to the summation of the two 2D spectra described above. Let us try addition or P-type peaks detection first. The FID detected in quadrature during t_2 after the 90°_x mixing pulse (the first experiment) is stored as usual in the computer memory. Channel A data (the x' component of the signal) represent the real part of the FID, $\mathrm{Re}(s(t_2))$; the data from channel B (the y' component) are stored as the imaginary part of the FID, $\mathrm{Im}(s(t_2))$. Such data routing corresponds to the receiver phase being set to $0°$ or x phase. Changing the phase of the mixing pulse by $90°$ to make it a 90°_x pulse (the second experiment) would normally also require changing the receiver phase by $90°$ in order to have the data properly added for time averaging. After the 90°_y pulse the x' pulse the x' component (channel A) has the same dependence on t_2 as the y' component has after the 90°_x pulse; it is the imaginary part of the FID and accordingly should be added to the imaginary part of the FID time averaging. The other component, which is detected in the B channel (the y' component), has the same time dependence as has been stored in the real part of the FID and hence should be added to it. For P-type detection, however, we want to add the real part of the FID obtained in the second scan to the imaginary part of the FID obtained in the first scan, and vice versa, in agreement with the addition of the spectra from the two experiments. This data treatment corresponds to having the receiver phase the same in the two scans. Hence the simplest form of P-type quadrature detection is performed if the following phase cycling is used.

Scan	First Pulse	Second Pulse	Receiver
1	x	x	x
2	x	y	x

This simplest phase cycling distinguishes the frequency signs but leaves unwanted axial peaks in the 2D spectrum. Let us recall that axial peaks are due to the magnetization that was directed along the z axis at time t_1. The second pulse has turned it into observable magnetization in the x', y' plane. Since a $180°$ phase shift of the second pulse inverts the intensity of the axial peaks (and does not affect the desired peaks), the axial peaks are easily eliminated by repeating the foregoing phase cycling once again but now with the second pulse having its phase incremented by $180°$. The phase cycling for P- or anti-echo detection is given in Table A.2.

Table A.2. Phase Cycling for Anti-echo Detection in COSY Spectra

Scan	First Pulse	Second Pulse	Receiver
1	x	x	x
2	x	$-x$	x
3	x	y	x
4	x	$-y$	x

Table A.3. Phase Cycling for Echo Detection in COSY Spectra

Scan	First Pulse	Second Pulse	Receiver
1	x	x	x
2	x	$-x$	x
3	x	y	$-x$
4	x	$-y$	x

Analogously, we can derive the phase cycling necessary for N-type or echo detection as shown in Table A.3.

The spectra obtained by the first Fourier transform from the FIDs acquired with the phase cycling described above vary with time t_1. The simple amplitude modulation is replaced, however, by phase or by combined phase and amplitude modulation. An example of phase modulation is shown in Fig. A.14. Obviously, magnetization with different frequency signs produces different modulations.

Both types of detection require four-step phase cycling; that is, the number of scans (FIDs) taken for time averaging for each value of time t_1 should be a multiple of 4. The elimination of artifacts during detection requires a CYCLOPS receiver and transmitter phase cycling, which should be superimposed on one of the cycling schemes above and which leads to a 16-step cycle.

Although the signal-to-noise ratio after 16 scans is often insufficient when signals from nuclei other than protons are time averaged, the need for having the number of scans a multiple of 16 often represents an intolerable increase in measuring time in the case of ^{1}H NMR. The experiment can be shortened fourfold if a pulsed magnetic field gradient is used instead of phase cycling to distinguish the direction of rotation during the evolution period. Two identical magnetic field gradient pulses just before and after the mixing pulse ensure that only N-peaks are detected. P-peaks are detected when the second gradient pulse has the same magnitude and duration as the first one but has inverted polarity.

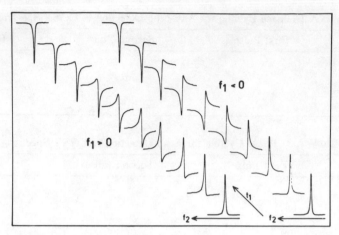

Figure A.14 Simulated phase modulation of a spectral line: left, modulation due to positive frequency; right, modulation due to negative frequency.

For clarity we have considered the simple case of a spectrum containing only diagonal peaks of isolated I nuclei. Our arguments can be straightforwardly extended to all diagonal peaks, since all originate from magnetizations that are left in the x', y' plane by the mixing pulse. All have the same phase. Cross-peaks, however, are caused by magnetization that has been turned from the x', y' plane to the $\pm z$ axis. Our discussion about frequency sign discrimination is equally valid for these peaks, but the cross-peaks, because of their different origin, have different phases. When the 2D spectrum is phased so that the cross-peaks appear as absorption peaks, the diagonal peaks are 90° out of phase.

All the data-handling procedures above yield 2D spectra with phase-twisted lineshapes. The route to pure absorption lineshapes [99] is suggested by the symmetry of the 2D spectra obtained in the two experiments discussed earlier in this appendix (see the spectra in Fig. A.12). For example, the dispersion lineshape can be removed from the spectrum measured with a 90°_{y} mixing pulse (the second experiment, spectrum b in Fig. A.12) by averaging the intensities in the points symmetrically located around the axis $f_1 = 0$. A similar symmetrization procedure can be applied to the spectrum obtained in the first experiment. The results, two 2D spectra which both have a pure absorption phase but without frequency sign discrimination, are shown in Fig. A.15. Since the symmetrization leaves one peak with inverted intensity, the two spectra can again be combined to achieve not only a pure absorption lineshape but also frequency discrimination, as indicated in Fig. A.15.

When a pure absorption lineshape is needed, the two experiments cannot

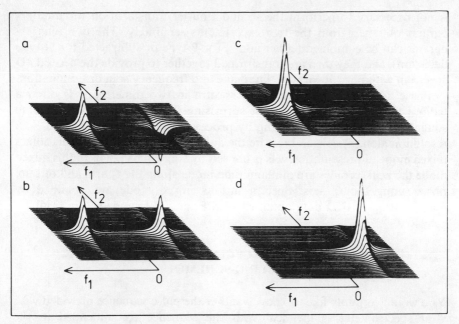

Figure A.15 Simulated 2D spectra with pure absorption lineshapes: (*a*) spectrum from the first experiment; (*b*) spectrum from the second experiment; (*c*) P-peak detection; (*d*) N-peak detection. The spectra can be obtained either by symmetrization or by zeroing the imaginary part before the second Fourier transform.

be combined during data accumulation; it is necessary to keep the results of the two experiments separate until almost the last processing stage.

The data handling for obtaining a pure absorption lineshape depends on the type of Fourier transform available in the spectrometer software. We shall look briefly into the processing employed with a complex Fourier transform (the method of States et al. [102]), since that appears more transparent. The method required on spectrometers equipped with real (or cosine) Fourier transform only is known as the time-proportional phase incrementation (TPPI) method, is described in the literature [99, 101], and should also be covered in the spectrometer manual.

According to States et al. [102], the pure absorption spectra are obtained from the two experiments, not as described above (i.e., by symmetrization) but by zeroing the imaginary part of the $s(t_1, f_2)$ signal after the first Fourier transform. That is, the FIDs are transformed and phase corrected, and then their imaginary parts are all set equal to zero. It can be shown mathematically [101, 102] that the second Fourier transform of a signal modified in such a way yield a pure absorption spectrum such as those shown in Fig. A.15. It is not necessary to perform the second Fourier transform on the interferograms obtained from the two experiments separately. The two interferograms can be combined (either added for P-type or subtracted for N-type detection), and they then are transformed together to provide the desired 2D spectrum with pure absorption lineshape and frequency spin discrimination.

Phase adjustment of pure phase spectra in two dimensions is often a tedious task. (Therefore, it is not surprising that its automation is being sought [283].) To avoid this lengthy process but preserve the needed high resolution along f_1 axis and restore the automation in spectral calculation, a mixed mode of presentation was proposed by Nagayama [284]. In this mixed mode the peak profiles are of magnitude mode along the f_2 axis and of pure phase along the f_1 axis. Spectra in this mixed mode are obtained by manipulation of FID data obtained with P- and N-type detection [284].

A.8. MORE ON PHASE CYCLING; A COMPARISON OF CYCLING SCHEMES

You would certainly like to know whether the pulse sequence provided with your spectrometer is identical with the sequence recommended in the literature for your intended purpose. To compare the number of pulses and the timing, the flip angles, and the delays between them is easy. It is more difficult to compare the phase cycling, but it is crucial. Different delays or flip angles lead only to a less-than-optimum performance, but a different phase cycling scheme produces completely different results. Thus, depending on the phase cycling used [187], a homonuclear pulse sequence consisting of three pulses can produce NOESY, RELAY, double quantum filtered COSY, or double quantum correlated spectra.

It is unfortunately likely that two sequences to be compared will have their phase cycling schemes described in different nomenclatures. In addition to giving the phases of transmitters and receiver as a relative angle expressed either in degrees or in radians (e.g., 30° or $\pi/12$), the most common phases of 0°, 90°, 180°, and 270° are also frequently specified as x, y, $-x$, and $-y$, or even as 0, 1, 2, and 3. Additionally, the easily understood tables of phase cycling are all too often replaced either by recurrent formulas or by some shorthand forms. For example, the CYCLOPS phase cycling for canceling hardware imperfections during quadrature detection might be written in shorthand as "00 11 22 33," where the phases for all pulses and the receiver are given in the order that the events occur in the sequence, with successive scans being separated by spaces. Two sequences claimed to serve the same purpose might still appear different even after the two sequences are placed on the same nomenclature basis. One should realize, however, that the phases have only a relative meaning; it was our arbitrary choice to designate one particular axis of the rotating coordinate system to be our x' axis. If we should label some other axis as the x' axis, nothing would change. These results are summarized in Patt's first rule [285]: "In any given experiment, the phases of all events in the sequence (transmitter, decoupler, and receiver phases) may be changed by adding a constant to all phases with no effect on the spin system." Thus time averaging in a simple one-pulsed sequence can be performed by repeating the sequence 00 00 00 00 00... (transmitter and receiver phases) until the desired signal-to-noise ratio is achieved. By adding 1 to both transmitter and receiver phases on successive scans, we get the CYCLOPS cycles 00 11 22 33 00, which produce the same result but with the additional bonus of eliminating imperfections from quadrature detection.

In designing a pulse cycling scheme that must perform several tasks (e.g., the schemes given in the preceding section had to detect either P- or N-peaks and eliminate the axial peaks), it is not important which of the tasks is performed earlier and which of the tasks later (we can eliminate the axial peaks and then detect N-peaks, or vice versa). Hence Patt's second rule: "In any experiment, we may rearrange the order of scans without affecting the result" [285].

To demonstrate the use of these rules, we shall compare the phase cycling for P-peak detection in the Jeener experiment given in the preceding section with that used by Keeler and Neuhaus [99]. In the shorthand above, the table for P-peak detection (Table A.3) can be written as

$$000 \ 010 \ 020 \ 030$$

Keeler and Neuhaus [99] replaced the phase shifting of the mixing pulse by phase shifts of the receiver and the pulse preceding the t_1 time:

$$000 \ 101 \ 202 \ 303$$

At first sight the two cycling schemes appear very different. Let us follow step by step the cycling scheme of Keeler and Neuhaus. The phase combinations for the first scan are the same in the two schemes. If we add the constant 3 to all phases in the second scan of Keeler and Neuhaus (first rule):

$$
\begin{array}{c}
101 \\
333 \\
\hline
\end{array}
$$

we obtain: 030

(In such additions there is no carry-over from one pulse to another; the addition is performed in base 4). The sum 030 is identical with the phase combination in our fourth scan and according to the second rule should not change the results. If we add 222 to the third-phase combination of Keeler and Neuhaus, we get 020, which is identical to our third combination. Summarizing, the two-phase cycling schemes are identical.

Your can similarly compare more complicated pulse sequences, although the procedure is of course tedious for extensive cycling schemes.

A.9. EXPERIMENTAL DETAILS

Nearly all the spectra used to illustrate the discussion in this book were measured on a Varian XL-200 spectrometer. It should be noted that the spectrometer hardware has not bee upgraded since its installation in 1979, and that the spectrometer computer is the older V77-220 data system, for which the last software version (H.2) was released in 1983, rather than the 68000 data system available with newer spectrometers of the same type. It was therefore necessary for the authors to program some pulse sequences or to obtain them from colleagues (to whom we are very grateful).

The requirements of 2D NMR spectroscopy (e.g., pulse and receiver phase shifts, small-angle digital phase shifters, additional computer control over spectrometer functions, larger CPU memory and disks, faster data handling, etc.) has notably affected the development of new commercial NMR spectrometer systems in the last few years (since 1982), and the use of the powerful 2D NMR technique has now become easier. Modern spectrometers are usually equipped by the manufacturer with a software package that includes the programs necessary for 2D data treatment and the most common of the pulse sequences.

The spectrometer used in this work was operated at frequencies of 200 and 50.3 MHz for ^{1}H and ^{13}C NMR measurements, respectively. Sample tubes with an outside diameter of 5 mm were used for all measurements (including 2D INADEQUATE). The 90° pulse length of nonselective pulses varied between 7.5 and 9.5 μs in the observation channel (^{1}H and ^{13}C NMR) and between 45 and 60 μs in the decoupler channel. The sample temperature, which was not controlled except for exchange spectra, varied from 19 to 24°C.

All significant 2D NMR methods were demonstrated on three samples of 2-butanol (Lachema, p.a. grade):

1. In deuteriochloroform, a 75% (v/v) solution shows the OH signal as a single line between the CH and CH_2 proton multiplets (Fig. A.16a).
2. In hexadeuteriobenzene, a 90% (v/v) solution also shows the OH

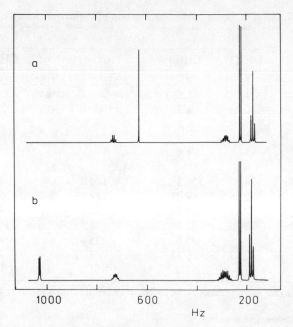

Figure A.16 ¹H NMR spectra of 2-butanol referenced to external TMS: (*a*) solution in deuteriochloroform; (*b*) dry solution in hexadeuteriobenzene.

proton signal as a broad singlet but at lower field (higher frequency) than the CH proton multiplet.

3. In hexadeuteriobenzene, a 75% (v/v) solution [i.e., the same as solution 2 but dried with molecular sieve (Fisher 5A)]. The OH proton signal appears as a doublet (Fig. A.16*b*).

Spectra of samples 1 and 3 are shown in Fig. A.16, with tetramethylsilane (TMS) as external reference. Spectra shown in other sections of the text were not referenced. Some significant NMR parameters of the 2-butanol used as an example throughout this text are collected in Table. A.4. Spectral width and spectrometer frequencies were chosen for data matrix economy. In most cases, the spectral width in the ¹H axis was 680 Hz (for samples 1 and 2) or 1050 Hz (for sample 3); in the ¹³C direction a spectral width of 3200 Hz was sufficient for decoupled spectra, while a larger width (up to 4500 Hz) was needed for coupled spectra (Fig. A.17).

Only some features of the three samples above need be considered if the reader wishes to reproduce the 2D NMR spectra shown in this book. The sample concentration can be lowered, especially if a strong lock signal is

Table A.4. Some NMR Parameters of 2-Butanol Samples

Parameter	CH$_3$(1)	CH(2)	CH$_2$(3)		CH$_3$(4)	OH
			Group[a]			
$\delta(^1H)^{b,c}$	1.161	3.670	1.533^d	1.418^d	0.927	5.165^e
$\delta(^1H)^{c,f}$	1.153	3.683	1.499^d	1.384^d	0.903	3.185^e
$^3J(^1H-^1H)^{b,g}$	6.2	6.2	7.4^d	7.4^d	7.4	—
	—	6.7	6.7	—	—	—
	—	5.8	—	5.8	—	—
	—	4.8^e	—	—	—	4.8^e
$^2J(^1H-^1H)^{b,g}$	—	—	−13.5	−13.5	—	—
$^1J(^{13}C-^1H)^{b,g}$	124.9	139.5	124.0		124.6	—
$\delta(^{13}C)^{b,c}$	22.98	69.01	32.31		10.32	—
$\delta(^{13}C)^{c,f}$	22.13	68.44	34.44		9.49	—
$^1J(^{13}C-^{13}C)^{b,g}$	38.7	38.7	34.8		34.8	—
	—	35.6	35.6		—	—
$T_1(^{13}C)^{b,h}$	2.32	3.24	2.32		4.24	—
$T_1(^1H)^{b,h,i}$	1.1	1.6	1.2		1.4	1.0

[a] The group number in parentheses is the carbon number according to the IUPAC numbering system.
[b] In hexadeuteriobenzene solution.
[c] Chemical shift in δ-scale.
[d] Diastereotopic protons; 500-MHz spectrum analysis.
[e] Concentration, temperature, and moisture dependent.
[f] Deuteriochloroform solution.
[g] Coupling constants in hertz units.
[h] Relaxation time in seconds.
[i] The longest relaxation time in the multiplet.

needed for spectrometer stabilization, and the exact position of the OH signal is unimportant as long as the signal remains between the CH and CH$_2$ multiplets in the spectrum of sample 1 above. Sample 1 should be used when it is desirable to reduce the spectral width along the ^1H axis, when coupling with the OH proton would complicate the spectrum, or when suppression of a singlet line has to be demonstrated. The difference between samples 2 and 3 is significant only for heteronuclear J-resolved spectra selectively resolved according to long-range coupling (the spectra of sample 2 would be much simpler than those shown).

Certain features of 2D NMR spectra were variously demonstrated on a furan derivative, acetone, chloroform, and on N,N-dimethylacetamide, since some NMR topics were more obvious with these compounds. Details about the solutions of these samples are given in the respective figures. All the samples are commercially available except the furan derivative, which could

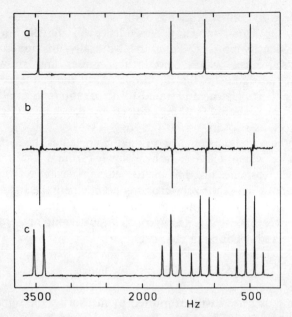

Figure A.17 ^{13}C NMR spectra of dry solution of 2-butanol in hexadeuteriobenzene referenced to external TMS: (*a*) decoupled spectrum; (*b*) 1D INADEQUATE spectrum; (*c*) coupled spectrum.

be replaced by any compound having a simple AX proton system. The chloroform was measured using a capillary filled with D_2O to provide the signal for the lock system.

A.10. PULSE FLIP ANGLE CALIBRATION

An exact setting of the pulse flip angles is not very critical for a satisfactory performance of most 2D NMR methods, although an incorrect setting of the pulse length always leads to reduced sensitivity. In such a case a pulse calibration for one sample (usually one with high concentration) can be utilized in the measurements of other samples in the same solvent (provided that the spectrometer probe is kept correctly tuned). Changing the solvent can alter the flip angles by 10% or more, especially when going from an organic solvent to ionic solutions, and vice versa. When such a misadjustment cannot be tolerated (e.g., in multiple quantum filtered or multiplicity edited experiments), the pulses must be calibrated directly on the sample under investigation.

In principle, any pulse sequence can be used for pulse calibration, provided that the signal intensity dependence on the flip angle can be predicted from density matrix calculations. Naturally, it is preferable to have simple pulse sequences which contain few pulses and show a simple dependence of the signal on the flip angle. Also, the sequence should be sufficiently sensitive to give a good signal-to-noise ratio for a small number of scans, the accuracy should not be affected by delay mismatches or off-resonance effects, and so on.

In general, the signal dependence on the flip angle has the form of sine or cosine functions which change rather slowly around the maxima and minima; thus the extremes are difficult to determine exactly. Therefore, it is more unambiguous to use the zero-crossing point where the signal is positive on one side and negative on the other side of the zero point. For heteronuclear experiments it is necessary to know the flip angles of pulses in both observing and decoupling channels.

A.10.1. Observing Channel

The observing pulses are easily calibrated by measuring the signal after one pulse (pulse sequence of Fig. 2.1). When the pulse width t_p is varied, the flip angle α is altered according to Eq. (A.3-2), and since the signal observed after one pulse sequence is

$$I = I_{max} \sin \alpha \qquad \text{(A.10-1)}$$

one can search for the pulse width corresponding to a maximum signal ($\alpha = 90°$) or a null signal ($\alpha = 180°$ or $360°$). For the reasons explained above, the determination of the pulse width of $180°$ or $360°$ pulses is more accurate, but it requires a good signal-to-noise ratio. In such a case, Eq. (A.3-2) is assumed to be valid and the pulse widths (t_p) for different flip angles can be calculated [282].

The assumption of proportionality between the flip angle α and the pulse width t_p is not necessarily valid on all spectrometers for all pulse widths. For short pulses the pulse distortions might have relatively large effects, and the true $90°$ pulse would then have a pulse width larger than one-half of a $180°$ pulse. In that situation it is better to use not one pulse but two identical pulses, with one closely following the other, and to search for the pulse width when the combined effect of the two pulses equals that of a $180°$ pulse (for details and solutions of possible technical problems, see ref. 5). Similarly, other pulses can be calibrated by repeating the pulse an appropriate number of times so that the combined effect of a $m\alpha$ pulses is that of a $180°$ or $360°$ pulse [5].

The search for the required pulse width can be shortened [286] by measuring intensities for two properly chosen pulse widths. If $t_{p1} = 2t_{p2}$, the ratio of the two intensities (I_1 and I_2) is

$$I_1/I_2 = 1/(2 \cos \alpha_1) \qquad (A.10\text{-}2)$$

According to Eq. (A.3-2), the pulse width $t_p(\alpha)$ of the pulse with the flip angle α is

$$t_p(\alpha) = (\alpha t_{p1})/\alpha_1 \qquad (A.10\text{-}3)$$

Thus the appropriate pulse width can be calculated from the two measurements of signal intensity.

The described calibration procedures require a relaxation delay of the order of $5 \times T_1$ in order to reestablish Boltzmann equilibrium before the pulse can be repeated. This can prove to be time consuming in the case of "rare" nuclei, which tend to have weak signals and long relaxation times. Wesener and Günther [287] have shown how the pulse width can be determined accurately under nonequilibrium conditions provided that a steady state between excitation and relaxation is reached.

If a steady state is established by high pulse repetition in the presence of spin-lattice relaxation, Eq. (A.10-1) must be modified to the form

$$I = I_{\max} \sin(\alpha \cdot S) \qquad (A.10\text{-}4)$$

where the steady-state factor S is

$$S = [1 - \exp(-\tau/T_1)]/[1 - \exp(-\tau/T_1) \cos \alpha] \qquad (A.10\text{-}5)$$

and τ is the delay between the two pulses or the pulse repetition time which satisfies the conditions $\tau \gg T_2$. It is clear from Eq. (A.10-4) that the search for a 90° pulse must give erroneous flip angles (the signal maximum is shifted from $\alpha = 90°$), and the search for 180° is more difficult (since the dependence is flattened around this angle) under nonequilibrium conditions. It is only the search for a 360° pulse that is still applicable for calibration. The authors suggest, however, a two-dimensional approach. If the pulse width is varied with evolution time, the f_1 frequency obtained after FT can then be used to calculate the pulse width for the required flip angle according to the relationship

$$t_p = \alpha/(2\pi f_1) \qquad (A.10\text{-}6)$$

and the overall time needed for pulse calibration is considerably shortened.

A.10.2.   Decoupling Channel

Calibration of the decoupler rf field (B_1) for continuous decoupling is well documented in NMR texts [281, 282] and spectrometer manuals. The calibration is based on measurements of a series of coherently off-resonance decoupled spectra in which the observed residual coupling or splitting (J_r, in hertz) is related to the true coupling constant value J (in hertz)

$$J_r = [(\Delta f - J/2)^2 + (\gamma B_1)^2]^{0.5} - [(\Delta f + J/2)^2 + (\gamma B_1)^2]^{0.5} \qquad \text{(A.10-7)}$$

where Δf is the decoupler offset from the resonance.

Although the pulse flip angle can be estimated from the value of B_1 determined from coherent continuous decoupling according to Eqs. (A.3-2) and (A.10-7), it is safer to use direct decoupler pulse calibration.

As already mentioned, any heteronuclear pulse sequence involving decoupler pulses can be used for calibration; very often, INEPT or DEPT sequences are used routinely for rough calibration (simply optimizing the decoupler pulse for the best polarization transfer [14]), but there are a few pulse sequences that have been specially designed for this purpose.

The simplest suggested sequence consists of two pulses [288–290],

$$I(\text{sat}) - 90°(S) - t - \theta(I) - \text{detect}(S) \qquad \text{(A.10-8)}$$

where $I(\text{sat})$ denotes continuous irradiation of the decoupled spins (protons) to establish NOE so that the detected signal will be enhanced and thus the sensitivity of the method will be increased. The signal of nuclei S is null at $\theta = 90°$ for all spin systems I_nS; for proper functioning of the sequence the time delay should be set $t = 1/(2 \cdot J_{IS})$. Various aspects of the use of this sequence have been considered by several authors [291–293]. In short, the simplest method has poor sensitivity since the signal must be detected without decoupling. The instrumental delay between the last pulse and the beginning of data acquisition causes phasing problems that lead to difficulty in finding the zero-crossing point.

Using the SEMUT pulse sequence [294] for calibration of the decoupler pulses

$$I(\text{sat}) - 90°(S) - t - \theta(I),180°(S) - t - \text{decouple}(I),\text{detect}(S) \qquad \text{(A.10-9)}$$

has the advantage that it allows decoupling during detection, and the phasing problems are eliminated by the refocusing 180°(S) pulse [291]. The sequence is also less sensitive to missetting the delay [291]; missetting for an IS spin system leads to estimates of decoupling fields that are lower than the true values [292]. The sequence has an intensity zero-crossing point at $\theta = 90°$ only for IS spin systems; for I_2S systems there is a minimum intensity at $\theta = 90°$ but no zero crossing [291, 292]; for I_3S systems the signal varies very slowly around $\theta - 90°$, although it does change sign. Hence only IS spin systems are suitable for calibration by SEMUT [as well as by pulse sequence (A.10-9); for more details, see ref. 292.

The method based on refocused SINEPT [295, 296] is more universally applicable.

$$\text{I:} \quad \theta - t - 2 \cdot \theta - t' - \text{decouple}$$
$$\text{S:} \qquad\quad 90° \qquad\qquad 180° - t' - \text{detect} \qquad \text{(A.10-10)}$$

It employs polarization transfer (and hence is more sensitive than the methods above in the case of nuclei with low magnetogyric ratios). The refocused SINEPT exhibits identical zero-crossing characteristics (at $\theta = 90°$) for all I_nS spin systems; it is the most accurate of the three methods described. Optimum sensitivity requires $t = t_1 = 1/(2 \cdot J_{IS})$, but zero crossing is independent of the delay matching.

The pulse sequence proposed by Sklenář et al. [293] has similar features:

$$S(\text{sat}), \theta_x(I) - t - m \cdot \theta_y(I), 180°(S) - \text{detect(s)} \qquad \text{(A.10-11)}$$

although an additional delay with a refocusing pulse would permit detection under decoupled conditions as discussed above. The saturation of observed nuclei, $S(\text{sat})$, can be accomplished by a train of closely spaced $90°(S)$ pulses in order to destroy natural S spin polarization. The zero-crossing points are at $0 - k \cdot 180°/m$, where $K = 0, 1, 2, \ldots$ for all I_nS spin systems. The larger the value of m employed, the steeper is the intensity dependence around the zero-crossing point.

2D variants of the foregoing 1D techniques of SEMUT, DEPT, INEPT, and SINEPT were designed by Sørensen et al. [196, 297]. In the 2D calibrating experiments, the flip angle of a properly chosen decoupler pulse in the sequence is incremented in the same way as the evolution time in other 2D experiments. The resulting 2D spectra exhibit splittings along the f_1 axis with the line separations proportional to the decoupling field (for the details of the analysis, see the original literature). The 2D techniques are superior to the parent 1D methods with respect to both sensitivity and accuracy; in ^{13}C NMR spectroscopy they offer significant time savings not only because of their higher sensitivity but also because they simultaneously provide information about the multiplicity of the lines in the spectrum.

REFERENCES

1. Bax, A. *Two-Dimensional Nuclear Magnetic Resonance in Liquids*, Delft University Press, Dordrecht, The Netherlands, 1982.
2. Ernst, R. R., Bodenhausen, G., Wokaun, A. *Principles of Nuclear Magnetic Resonance in One and Two Dimensions*, Clarendon Press, Oxford, 1987.
3. Harris, R. K. *Nuclear Magnetic Resonance Spectroscopy, A Physicochemical View*, Pitman, Marshfield, Mass., 1983; Halsted (Wiley), New York, 1986.
4. Farrar, T. C., Becker, E. D. *Pulse and Fourier Transform NMR, Introduction to Theory and Methods*, Academic Press, New York, 1971.
5. Fukushima, E., Roeder, S. B. W. *Experimental Pulse NMR, a Nuts and Bolts Approach*, Addison-Wesley, Reading, Mass., 1981.
6. Allerhand, A., Addleman, R. E., Osman, D. *J. Am. Chem. Soc. 107*, 5809 (1985).
7. Hull, W. E., Wehrli, F. W. in *Topics in Carbon-13 NMR Spectroscopy* (G. C. Levy, Ed.), Vol. 4, p. 1, Wiley-Interscience, New york, 1984.
8. Morris, G. A. in *Fourier Transforms in Chemistry* (A. G. Marshall, Ed.), p. 271, Plenum Press, New York, 1982.
9. Jeener, J. Ampère International Summer School II, Basko Polje, Yugoslavia, 1971, unpublished.
10. Aue, W. P., Bartholdi, E., Ernst, R. R. *J. Chem. Phys. 64*, 2229 (1976).
11. Benn, R., Günther, H. *Angew. Chem. Int. Ed. Engl. 22*, 350 (1983).
12. Freeman, R., Morris, G. A. *Bull. Magn. Reson. 1*, 5 (1979).
13. Freeman, R. *Proc. R. Soc. London A373*, 149 (1980).
14. Morris, G. A. *Magn. Reson. Chem. 24*, 371 (1986).
15. Bax, A. in *Topics in Carbon-13 NMR Spectroscopy* (G. C. Levy, Ed.), Vol. 4, p. 197, Wiley-Interscience, New York, 1984.
16. Turner, D. L. *Prog. NMR Spectrosc. 17*, 281 (1985).
17. Turner, C. J. *Prog. NMR Spectrosc. 16*, 311 (1984).
18. Croasmun, W. R., Carlson, R., Eds. *Methods in Stereochemical Analysis: Two-Dimensional NMR Spectroscopy: Applications for Chemists and Biochemists* VCH Publishers, Deerfield Beach, Fla., 1987.
19. Brey, W. (Ed.) *Pulse Methods in 1D and 2D Liquid Phase NMR*, Academic Press, Orlando, Fla. 1987.
20. Gray, G. A. in ref. 18.

21. Gray, G. A. in ref. 19.
22. Hull, W. E. in ref. 18.
23. Bernstein, M. A. in ref. 18.
24. Dabrowski, J. in ref. 18.
25. Nagayama, K. in *NMR in Stereochemical Analysis* (Takechi, Y., Marchand, A. P., Eds.), Vol. 6, Chap. 5, VCH Publishers Inc., Deerfield Beach, Fla., 1986.
26. Kessler, H., Bermel, W. in *NMR in Stereochemical Analysis* (Takeuchi, Y., Marchand, A. P., Eds.), Vol. 6, Chap. 6, VCH Publishers Inc., Deerfield Beach, Fla., 1986.
27. Blümich, B., Ziessow, D. *J. Magn. Reson. 49*, 151 (1982).
28. Müller, L., Kumar, A., Ernst, R. R. *J. Chem. Phys. 63*, 5490 (1975).
29. Maudsley, A. A., Ernst, R. R. *Chem. Phys. Lett. 50*, 368 (1977).
30. Jakobsen, H. J., Linde, S. AA., Sørensen, S. *J. Magn. Reson. 15*, 385 (1974).
31. Pachler, K. G. R., Wessels, P. L. *J. Magn. Reson. 12*, 337 (1973).
32. Bodenhausen, G., Freeman, R. *J. Magn. Reson. 28*, 471 (1977).
33. Levitt, M. H., Bodenhausen, G., Ernst, R. R. *J. Magn. Reson. 58*, 462 (1984).
34. Aue, W. P., Bachmann, P., Wokaun, A., Ernst, R. R. J. *Magn. Reson. 29*, 523 (1978).
35. Lindon, J. C., Ferrige, A. G., *Prog. NMR Spectrosc. 14*, 27 (1980).
36. Denk, W., Wagner, G., Rance, M., Wüthrich, K. *J. Magn. Reson. 62*, 350 (1985).
37. Abraham, R. J. *The Analysis of High-Resolution NMR Spectra*, Elsevier, Amsterdam, 1971.
38. Corio, P. L. *Structure of High-Resolution NMR Spectra*, Academic Press, New York, 1966.
39. Bodenhausen, G., Freeman, R., Niedermeyer, R., Turner, D. L. *J. Magn. Reson. 24*, 291 (1976).
40. Freeman, R., Hill, H. D. W. in *Dynamic Nuclear Magnetic Resonance Spectroscopy* (L. M. Jackman, F. A. Cotton, Eds.) p. 131, Academic Press, New York, 1975.
41. Bodenhausen, G., Freeman, R., Turner, D. L. *J. Chem. Phys. 65*, 839 (1976).
42. Carr, H. Y., Purcell, E. M. *Phys. Rev. 94*, 630 (1954).
43. Müller, L., Kumar, A., Ernst, R. R. *J. Magn. Reson. 25*, 383 (1977).
44. Bodenhausen, G., Freeman, R., Niedermeyer, R., Turner, D. L. *J. Magn. Reson. 26*, 133 (1977).
45. Bodenhausen, G., Freeman, R., Turner, D. L. *J. Magn. Reson. 27*, 511 (1977).
46. Bodenhausen, G., Turner, D. L. *J. Magn. Reson. 41*, 200 (1980).
47. Freeman, R., Keller, J. *J. Magn. Reson. 43*, 484 (1981).
48. Hansen, M., Jakobsen, H. J. *J. Magn. Reson. 10*, 74 (1973).
49. Freeman, R., Kempsell, S. P., Levitt M. H. *J. Magn. Reson. 34*, 663 (1979).

50. Freeman, R., Morris, G. A., Turner, D. L. *J. Magn. Reson.* *26*, 373 (1977).

51. Bodenhausen, G., Freeman, R., Morris, G. A., Turner, D. L. *J. Magn. Reson.* *28*, 17 (1977).

52. Wang, J.-S., Zhao, D.-Z., Ji, T., Han, X.-W., Cheng, C.-B. *J. Magn. Reson.* *48*, 216 (1982).

53. Noggle, J. H., Schimer, R. E. *The Nuclear Overhauser Effect, Chemical Applications*, Academic Press, New York, 1971.

54. Morris, G. A., Freeman, R. *J. Am. Chem. Soc.* *101*, 760 (1979).

55. Morris, G. A. *J. Am. Chem. Soc.* 102, 428 (1980).

56. Burum, D. P., Ernst, R. R. *J. Magn. Reson.* *39*, 163 (1980).

57. Bolton, P. *J. Magn. Reson.* *41*, 287 (1980).

58. Doddrell, D. M. Pegg, D. T., Brooks, W., Bendall, M. R. *J. Am. Chem. Soc.* *103*, 727 (1981).

59. Doddrell, D. M., Pegg, D. T., Bendall, M. R. *J. Magn. Reson.* *48*, 323 (1982).

60. Pegg. D. T., Doddrell, D. M., Bendall, M. R. *J. Chem. Phys.* *77*, 2745 (1982).

61. Rutar, V. *J. Am. Chem. Soc.* *105*, 4095 (1983).

62. Davis, D. G., Agosta, W. C., Cowburn, D. *J. Am. Chem. Soc.* *105*, 6189 (1983).

63. Bax, A., Freeman, R. *J. Am. Chem. Soc.* *104*, 1099 (1982).

64. Turner, D. L. *J. Magn. Reson.* *39*, 391 (1980).

65. Bax, A. *J. Magn. Reson.* *52*, 330 (1983).

66. Rutar, V. *J. Magn. Reson.* *56*, 87 (1984).

67. Rutar, V. *J. Magn. Reson.* *58*, 132 (1984).

68. Aue, W. P., Karhan, J. Ernst, R. R. *J. Chem. Phys.* *64*, 4226 (1976).

69. Nagayama, K., Bachmann, P., Wüthrich, K., Ernst, R. R. *J. Magn. Reson.* *31*, 133 (1978).

70. Mersh, J. D., Sanders, J. K. M. *J. Magn. Reson.* *50*, 171 (1982)

71. Freeman, R., Hill, H. D. W. *J. Chem. Phys.* *54*, 301 (1971).

72. Shaka, A. J., Keeler, J., Freeman. R *J. Magn. Reson.* *56*, 294 (1984).

73. Bax, A., Mehlkopf, A. F., Smidt, J. *J. Magn. Reson.* *35*, 373 (1979).

74. Bax, A., Mehlkopf, A. F., Smidt, J. *J. Magn. Reson.* *40*, 213 (1980).

75. Macura, S., Brown, L. R. *J. Magn. Reson.* *53*, 529 (1983).

76. Nagayama, K. *J. Chem. Phys.* *71*, 4404 (1979).

77. Freeman, R., Kempsell, S. P., Levitt M. H. *J. Magn. Reson.* *34*, 663 (1979).

78. Williamson, M. P. *J. Magn. Reson.* *55*, 471 (1983).

79. Bodenhausen, G., Freeman, R., Morris, G. A., Turner, D. L. *J. Magn. Reson.* *31*, 75 (1978).

80. Wider, G., Baumann, R., Nagayama, K., Ernst, R. R., Wüthrich, K. *J. Magn. Reson.* *42*, 73 (1981).

81. Hall, L. D., Sukumar, S., Sullivan, G. R. *J. Chem. Soc. Chem. Comm.*, 292 (1979).

82. Nagayama, K., Wüthrich, K. Eur. *J. Biochem. 114*, 365 (1981).

83. Bleich, H., Gould, S., Pitner, P., Wilde, J. *J. Magn. Reson. 56*, 515 (1984).

84. Waterhouse, A. L., Holden, I., Casida, J. E. *J. Chem. Soc. Perkin Trans. II*, 1011 (1985).

85. Halterman, R. L., Nguyen, N. H., Vollhardt, K. P. C. *J. Am. Chem. Soc. 197*, 1379 (1985).

86. Somers, T. C., White, J. D., Lee, J. J., Keller, P. J., Chang, C., Floss, H. G. *J. Org. Chem. 51*, 464 (1986).

87. Sato, Y., Kohnert, R., Gould, S. J. *Tetrahedron Lett. 27*, 143 (1986).

88. Maudsley, A. A., Müller, L., Ernst, R. R. *J. Magn. Reson. 28*, 463 (1977).

89. Bax, A. *J. Magn. Reson. 53*, 517 (1983).

90. Rutar, V. *J. Magn. Reson. 58*, 306 (1984).

91. Rutar, V. *Chem. Phys. Lett. 106*, 258 (1984).

92. Wong, T. C., Rutar, V., Wang, L. S. J. Am. Chem. Soc. 106, 7046 (1984).

93. Wilde, J. A., Bolton, P. H. *J. Magn. Reson. 59*, 343 (1984).

94. Wong, T. C. *J. Magn. Reson. 63*, 179 (1985).

95. Reynolds, W. F., Hughes, D. W., Perpick-Dumont, M., Enriquez, R. G. *J. Magn. Reson. 64*, 304 (1985).

96. Nakashima, T. T., John, B. K., McClung, R. E. D. *J. Magn. Reson. 59*, 124 (1984).

97. Bax, A., Morris, G. A. *J. Magn. Reson. 42*, 501 (1981).

98. Maudsley, A. A., Wokaun, A., Ernst, R. R. *Chem. Phys. Lett. 55*, 9 (1978).

99. Keeler, J., Neuhaus, D. *J. Magn. Reson. 63*, 454 (1985).

100. Bolton, P. H., Bodenhausen, G. *J. Magn. Reson. 46*, 306 (1982).

101. Marion, D., Wüthrich, K. *Biochem. Biophys. Res. Commun. 113*, 967 (1983).

102. States, D. J., Haberkorn, R. A., Ruben, D. J. *J. Magn. Reson. 48*, 286 (1982).

103. Freeman, R., Morris, G. A. *J. Chem. Soc. Chem. Commun.*, 684 (1978).

104. Thomas, D. M., Bendall, M. R., Pegg, D. T., Doddrell, D. M., Field, J. *J. Magn. Reson. 42*, 298 (1981).

105. Rutar, V., Wong, T. C. *J. Magn. Reson. 53*, 495 (1983).

106. Bendall, M. R., Pegg, D. T. *J. Magn. Reson. 53*, 144 (1983).

107. Bendall, M. R., Pegg, D. T., Doddrell, D. M. *J. Magn. Reson. 45*, 8 (1981).

108. Pegg, D. T., Bendall, M. R. *J. Magn. Reson. 55*, 114 (1983).

109. Nakashima, T. T., John, B. K., McClung, R. E. D. *J. Magn. Reson. 57*, 149 (1984).

110. Levitt, M. H., Sørensen, O. W., Ernst, R. R. *Chem. Phys. Lett. 94*, 540 (1983).

111. Kessler, H., Griesinger, C., Zarbock, J., Loosli, H. R. *J. Magn. Reson. 57*, 331 (1984).

112. Reynolds, W. F., Hughes, D. W., Perpick-Dumont, M., Enriquez, R. G. *J. Magn. Reson. 63*, 413 (1985).

113. Reynolds, W. *Magn. Moments 1*, 9 (1985).

114. Bodenhausen, G. *J. Magn. Reson. 39*, 175 (1980).

115. Morris, G. A. *J. Magn. Reson. 44*, 277 (1981).

116. Bauer, C., Freeman, R., Wimperis, S. *J. Magn. Reson. 58*, 526 (1984).

117. Bax, A., Freeman, R. *J. Magn. Reson. 45*, 177 (1981).

118. Bodenhausen, G., Ruben, D. J. *J. Chem. Phys. Lett. 69*, 185 (1980).

119. Neuhaus, D., Keller, J., Freeman, R. *J. Magn. Reson. 61*, 553 (1985).

120. Müller, L. *J. Am. Chem. Soc. 101*, 4481 (1979).

121. Bax, A., Griffey, R. H., Hawkins, B. L. *J. Magn. Reson. 55*, 301 (1983).

122. Bax, A., Subramanian, S. *J. Magn. Reson. 67*, 565 (1986).

123. Bax, A., Griffey, R. H., Hawkins, B. L. *J. Am. Chem. Soc. 105*, 7188 (1983).

124. Live, D. H., Davis, D. G., Agosta, W. C., Cowburn, D. *J. Am. Chem. Soc. 106*, 6104 (1984).

125. Benn, R., Brevard, C. *J. Am. Chem. Soc. 108*, 5622 (1986).

126. Bax, A., Summers, M. F. *J. Am. Chem. Soc. 108*, 2093 (1986).

127. Bendall, M. R., Pegg, D. T., Doddrell, D. M. *J. Magn. Reson. 52*, 81 (1983).

128. Frey, M. Wagner, G., Vašák, M., Sørensen, O. W., Neuhaus, D., Wörgötter, E., Kägi, J. H. R., Ernst, R. R., Wüthrich, K. *J. Am. Chem. Soc. 197*, 6847 (1985).

129. Brühwiler, D., Wagner, G. *J. Magn. Reson 69*, 546 (1986).

130. Slenář, V., Bax, A. *J. Magn. Reson. 71*, 379 (1987).

131. Wörgötter, E., Wagner, G., Wüthrich, K. *J. Am. Chem. Soc. 108*, 6162 (1986).

132. Otting, G., Senn, H., Wagner, G., Wüthrich, K. *J. Magn. Reson. 70*, 500 (1986).

133. Bax, A., Freeman, R. *J. Magn. Reson. 44*, 542 (1981).

134. Bax, A., Freeman, R., Morris, G. *J. Magn. Reson. 42*, 164 (1981)

135. Kumar, A., Hosur, R. V., Chandrasekhar, K. *J. Magn. Reson. 60*, 143 (1984).

136. Mayor, S., Hosur, R. V. *Magn. Reson. Chem. 23*, 470 (1985).

137. Hosur, R. V., Chary, K. V. R., Kumar, A., Govil, G. *J. Magn. Reson. 62*, 123 (1985).

138. Kumar, A., Hosur, R. V., Chandrasekhar, K., Murali, N. *J. Magn. Reson. 63*, 107 (1985).

139. Rance, M., Wagner, G., Sørensen, O. W., Wüthrich, K., Ernst, R. R. *J. Magn. Reson. 59*, 250 (1984).

140. Bax, A., Mehlkopf, A. F., Smidt, J. *J. Magn. Reson. 35*, 167 (1979).

141. Brown, L. R. *J. Magn. Reson. 57*, 513 (1984).

142. Hosur, R.V., Kumar, M.R., Sheth, A. *J. Magn. Reson. 65*, 375 (1985).

143. Nagayama, K., Kumar, A., Wüthrich, K., Ernst, R. R. *J. Magn. Reson. 40*, 321 (1980).

144. Nagayama, K., Wüthrich, K., Ernst, R. R. *Biochem. Res. Commun. 90*, 305 (1979).

145. Braunschweiler, L., Ernst, R. R. *J. Magn. Reson. 53*, 521 (1983).

146. Chandrakumar, N., Subramanian, S. *J. Magn. Reson. 62*, 346 (1985).

147. Chandrakumar, N. *J. Magn. Reson. 71*, 322 (1987).

148. Hartmann, S.R., Hahn, E. L. *Phys. Rev. 128*, 2042 (1962).

149. Davis, D. G., Bax, A. *J. Am. Chem. Soc. 107*, 2820 (1985).

150. Bax, A., Davis, D. G., Sarkar, S. K. *J. Magn. Reson. 63*, 230 (1985).

151. Bax, A., Davis, D. G. *J. Magn. Reson. 65*, 355 (1985).

152. Ohuchi, M., Hosono, M., Matushita, K., Imanari, M. *J. Magn. Reson. 43*, 499 (1981).

153. Wider, G., Hosur, R. V., Wüthrich, K. *J. Magn. Reson. 52*, 130 (1983).

154. Hore, P. J. *J. Magn. Reson. 56*, 535 (1984).

155. Haasnoot, C. A. G., Hilbers, C. W. *Biopolymers 22*, 1259 (1983).

156. Guittet, E., Delsuc, M. A., Lellemand, J. Y. *J. Am. Chem. Soc. 106*, 4278 (1984).

157. Santos, H., Turner, D. L. Xavier, A. V. *J. Magn. Reson. 58*, 344 (1984).

158. Tong, J. P. K., Kotovych, G. *J. Magn. Reson. 69*, 511 (1986).

159. Kao, L.-F., Hruby, V. J. *J. Magn. Reson. 70*, 394 (1986).

160. Morris, G. A., Smith, K. I., Waterton, J. C. *J. Magn. Reson. 68*, 526 (1986).

161. Sørensen, O. W., Rance, M., Ernst, R. R. *J. Magn. Reson. 56*, 527 (1984).

162. Oschkinat, H., Pastore, A., Pfändler, P., Bodenhausen, G. *J. Magn. Reson. 69*, 559 (1986).

163. Bolton, P. H. *J. Magn. Reson. 48*, 336 (1982).

164. Kessler, H., Bermel, W., Griesinger, C. *J. Magn. Reson. 62*, 573 (1985).

165. Sarkar, S. K., Bax, A. *J. Magn. Reson. 63*, 512 (1985).

166. Bax, A. *J. Magn. Reson. 53*, 149 (1983).

167. Kessler, H., Bernd, M., Kogler, H., Zarbock, J., Sørensen, O. W., Bodenhausen, G., Ernst, R. R. *J. Am. Chem. Soc. 105*, 6944 (1983).

168. Sørensen, O. W., Ernst, R. R. *J. Magn. Reson. 55*, 338 (1983).

169. Field, L. D., Messerle, B. A. *J. Magn. Reson. 66*, 483 (1986).

170. Field, L. D., Messerle, B. A. *J. Magn. Reson. 62*, 453 (1985).

171. Bolton, P. H. *J. Magn. Reson. 62*, 143 (1985).

172. Lerner, L., Bax, A. *J. Magn. REson. 69*, 375 (1986).

173. Bolton, P. H. *J. Magn. Reson. 63*, 225 (1985).

174. Kogler, H., Sørensen, O.W., Bodenhausen, G., Ernst, R. R. *J. Magn. Reson. 55*, 157 (1983).

175. Bolton, P. H. *J. Magn. Reson. 54*, 333 (1983).

176. Bolton, P. H., Bodenhausen, G. *Chem. Phys. Lett. 89*, 139 (1982).

177. Homans, S. W., Dwek, R. A., Fernandes, D. L., Rademacher, T. W. *Proc. Nat. Acad. Sci. USA, 81*, 6286 (1984).

178. Eich, G., Bodenhausen, G., Ernst, R. R. *J. Am. Chem. Soc. 104*, 3731 (1982).

179. Bax, A., Drobny, G. *J. Magn. Reson. 61*, 306 (1985).

180. Wagner, G. *J. Magn. Reson.* *55*, 151 (1983).

181. Weber, P. L., Drobny, G., Reid, B. R. *Biochemistry* *24*, 4549 (1985).

182. Delsuc, M. A., Guittet, E., Trotin, N., Lallemand, J. Y. *J. Magn. Reson.* *56*, 163 (1984).

183. Neuhaus, D., Wider, G., Wagner, R., Wüthrich, K. *J. Magn. Reson.* *57*, 164 (1984).

184. Bodenhausen, G. *Prog. Nucl. Magn. Reson. Spectrosc.* *14*, 137 (1982).

185. Sørensen, O. W., Eich, G. W., Levitt, M. H., Bodenhausen, G., Ernst, R. R. *Prog. NMR Spectrosc.* *16*, 163 (1983).

186. Bain, A. D. *J. Magn. Reson.* *56*, 418 (1984).

187. Bodenhausen, G., Kogler, H., Ernst, R. R. *J. Magn. Reson.* *58*, 370 (1984).

188. Bax, A., Freeman, R., Kemsell, S. P. *J. Magn. Reson.* *41*, 349 (1980).

189. Bodenhausen, G., Dobson, C. M. *J. Magn. Reson.* *44*, 212 (1981).

190. Hore, P. J., Zuiderweg, E. R. P., Nicolay, K., Dijkstra, D., Kapstein, R. *J. Am. Chem. Soc.* *104*, 4286 (1982).

191. Piantini, U., Sørensen, O. W., Ernst, R. R. *J. Am. Chem. Soc.* *104*, 6800 (1982).

192. Dumoulin, C. L. *J. Magn. Reson.* *64*, 38 (1985).

193. Rance, M., Sørensen, O. W., Bodenhausen, G., Wagner, G., Ernst, R. R., Wüthrich, K. *Biochem. Biophys. Res. Commun.* *117*, 479 (1983).

194. Thomas, M. A., Kumar, A. *J. Magn. Reson.* *61*, 540 (1985).

195. Boyd, J., Redfield, C. *J. Magn. Reson.* *68*, 67 (1986).

196. Sørensen, O. W., Neilsen, N. C., Bildsøe, H., Jakobsen, H. J. *J. Magn. Reson.* *70*, 54 (1986).

197 Farmer, B. T., II, Brown, L. R. *J. Magn. Reson.* *71*, 365 (1987).

198. Kessler, H., Oschkinat, H., Sørensen, O. W., Kogler, H., Ernst, R. R. *J. Magn. Reson.* *55*, 329 (1983).

199. Bax, A., Freeman, R., Kempsell, S. P. *J. Am. Chem. Soc.* *102*, 4849 (1980).

200. Bax. A., Freeman, R., Frenkiel, T. A. *J. Am. Chem. Soc.* *103*, 2102 (1981).

201. Bax, A., Freeman. R., Frenkiel, T. A., Levitt, M. H. *J. Magn. Reson.* *43*, 478, (1981).

202. Mareci, T. H., Freeman, R. *J. Magn. Reson.* *48*, 158 (1982).

203. Piveteau, D., Delsuc, M. A., Guittet, E., Lallemand, J. Y. *Magn. Reson. Chem.* *23*, 127 (1985).

204. Richarz, R., Ammann, W., Wirthlin, T. *J. Magn. Reson.* *45*, 270 (1981).

205. Turner, D. L. *J. Magn. Reson.* *53*, 259 (1983).

206. Turner, D. L. *J. Magn. Reson.* *49*, 175 (1982).

207. Bax, A., Mareci, T. H. *J. Magn. Reson.* *53*, 360 (1983).

208. Turner, D. L. *J. Magn. Reson.* *58*, 500 (1984).

209. Bolton, P. H. *J. Magn. Reson.* *68*, 180 (1986).

210. Sørensen, O. W., Freeman., R., Frenkiel, T., Mareci, T. H., Schuck, R. *J. Magn. Reson. 46*, 180 (1982).

211. Sparks, S. W., Ellis, P. D. *J. Magn. Reson. 62*, 1 (1985).

212. Keller, P. J., Vogele, K. E. *J. Magn. Reson. 68*, 389 (1986).

213. Rance, M., Sørensen, O. W., Leupin, W., Kogler, H., Wüthrich, K., Ernst, R. R. *J. Magn. Reson. 61*, 67 (1985).

214. Sørensen, O. W., Sørensen, U. B., Jakobsen, H. J. *J. Magn. Reson. 59*, 332 (1984).

215. Dalvit, C., Rance, M., Wright, P. E., *J. Magn. Reson. 69*, 356 (1986).

216. Wagner, G., Zuiderweg, E. R. P. *Biochem. Biophys. Res. Commun. 113*, 854 (1983).

217. Boyd, J., Dobson, C. M., Redfield, C. *J. Magn. Reson. 55*, 170 (1983).

218. Rance, M., Wright, P. E. *J. magn. Reson. 66*, 372 (1986).

219. Hore, P. J., Scheek, R. M., Kaptein, R. *J. Magn. Reson. 52*, 339 (1983).

220. Bain, A. D., Hughes, D. W., Coddington, J. M., Bell, R. A. *J. Magn. Reson. 58*, 490 (1984).

221. Levitt, M. H., Ernst, R. R. *Mol. Phys. 50*, 1109 (1983).

222. Levitt, M. H. *Prog. NMR Spectrosc 18*, 61 (1986).

223. Levitt, M. H., Ernst. R. R. *J. Chem. Phys. 83*, 3297 (1985).

224. Wimperis, S., Bodenhausen, G. *J. Magn. Reson. 71*, 355 (1987).

225. Bax, A., Freeman, R. *J. Magn. Reson. 41*, 507 (1980).

226. Jeener, J., Meier, B. H., Bachmann, P., Ernst, R. R. *J. Chem. Phys. 71*, 4546 (1979).

227. Kumar, A., Ernst, R. R., Wüthrich, K. *Biochem. Biophys, Res. Commun. 95*, 1 (1980).

228. Kumar, A., Wagner, G., Ernst, R. R., Wüthrich, K. *Biochem. Biophys. Res. Commun. 96*, 1156 (1980).

229. Wagner, G., Kumar, A., Wüthrich, K. *Eur. J. Biochem. 114*, 375 (1981).

230. Jeener, J., Meier, B. H., Bachman, P., Ernst, R. R. *J. Chem. Phys. 71*, 4546 (1979).

231. Meier, B. H., Ernst, R. R. *J. Am. Chem. Soc. 101*, 6441 (1979).

232. Wider, G., Macura, S., Kumar, A., Ernst, R. R., Wüthrich, K. *J. Magn. Reson. 56*, 207 (1984).

233. Kumar, A., Ernst, R. R., Wüthrich, K. *Biochem. Biophys. Res. Commun. 95*, 1 (1980).

234. Bodenhausen, G., Ernst, R. R. *J. Am. Chem. Soc. 104*, 1304 (1982).

235. Macura, S., Huang, Y., Suter, D., Ernst. R. R. *J. Magn. Reson. 43*, 259 (1981).

236. Johnston, E. R., Dellwo, M. J., Hendrix, J. *J. Magn. Reson. 66*, 399 (1986).

237. Neuhaus, D., Wagner, G., Vašák, M., Kägi, J. H. R., Wüthrich, K. *Eur. J. Biochem. 151*, 257 (1985).

238. Macura, S. Ernst, R. R. *Mol. Phys., 41*, 95 (1980).

239. Rance, M., Bodenhausen, G., Wagner, G., Wüthrich, K., and Ernst, R. R. *J. Magn. Reson. 62*, 497 (1955).

240. Hennig, J., Limbach, H. H. *J. Magn. Reson. 49*, 322 (1982).

241. Macura, S., Wüthrich, K., Ernst, R. R. *J. Magn. Reson. 47*, 351 (1982).

242. Macura, S. Wüthrich, K., Ernst, R. R. *J. Magn. Reson. 46*, 269 (1982).

243. Bodenhausen, G., Ernst, R. R. *J. Am. Chem. Soc. 104*, 1304 (1982).

244. Bremer, J., Mendz, G. L., Moore, W. J. *J. Am. Chem. Soc. 106*, 4691 (1984).

245. Kumar, A., Ernst, R. R., Wüthrich, K. *Biochem. Biophys. Res. Commun. 95*, 1 (1980).

246. Kumar, A., Wagner, G., Ernst, R. R., Wüthrich, K. *J. Am. Chem. Soc. 103*, 3654 (1981).

247. Haasnoot, C. A. G., van de Ven, F. J. M., Hilbers, C. W. *J. Magn. Reson. 56*, 343 (1984).

248. Gurevich, A. Z., Barsukov, I. L., Arseniev, A. S., Bystrov, V. F. *J. Magn. Reson. 56*, 471 (1984).

249. Wagner, G. *J. Magn. Reson. 57*, 497 (1984).

250. Rinaldi, P. L. *J. Am. Chem. Soc. 105*, 5167 (1983).

251. Yu, C., Levy, G. C. *J. Am. Chem. Soc., 106*, 6533 (1984).

252. Bolton, P. H. *J. Magn. Reson. 70*, 344 (1986).

253. Meier, B. U., Bodenhausen, G., Ernst, R. R. *J. Magn. Reson. 60*, 161 (1984).

254. Pfändler, P., Bodenhausen, G., Meier, B. U., Ernst, R. R. *Anal. Chem. 57*, 2510 (1985).

255. Neidig, K. P., Bodenmueller, H., Kalbitzer, H. R. *Biochem. Biophys. Res. Commun. 125*, 1143 (1984).

256. Pfändler, P., Bodenhausen, G. *J. Magn. Reson. 70*, 71 (1986).

257. Bodenhausen, G. *J. Mol. Struct. 141*, 255 (1986).

258. Macura, S., Wüthrich, K., Ernst, R. R. *J. Magn. Reson. 47*, 351 (1982).

259. Kessler, H., Bermel, W., Müller, A., Pook, K.-H. *Peptides 7*, 437 (1985).

260. Champeney, D. C. *Fourier Transforms and Their Physical Applications*, Academic Press, London, 1973.

261. Unkefer, C. J., Earl, W. L. *J. Magn. Reson. 61*, 343 (1985).

262. Hoult, D. I. Prog. *NMR Spectrosc. 12*, 41 (1978).

263. Hoult, D. I., Richards, R. E. *Proc. R. Soc. London A344*, 311 (1975).

264. Stejskal, E. O., Schaefer J. *J. Magn. Reson. 14*, 160 (1974).

265. Levitt, M. H., Freeman, R. *J. Magn. Reson. 33*, 473 (1979).

266. Wimperis, S., Freeman, R. *J. Magn. Reson. 58*, 348 (1984).

267. Barker, P., Freeman, R. *J. Magn. Reson. 64*, 334 (1985).

268. Counsell, C. J. R., Levitt, M. H., Ernst, R. R. *J. Magn. Reson. 64*, 470 (1985).

269. Bax, A., Sarkar, S. K. *J. Magn. Reson. 60*, 170 (1984).

270. Hahn, E. L. *Phys. Rev. 80*, 580 (1950).

271. Morris, G. A. in *Topics in Carbon-13 NMR Spectroscopy* (G. C. Levy, Ed.), Vol. 4, p. 179, Wiley, New York 1984.

272. Blinka, T. A., Helmer, B. J., West, R. *Adv. Organomet. Chem. 23*, 193 (1984).

273. Pegg, D. T., Doddrell, D. M., Bendall, M. R. *Magn. Reson. 51*, 264 (1983).

274. Blechta, V., Schraml, J. *J. Magn. Reson. 69*, 293 (1986).

275. Garbow, J. R., Weitekamp, D. P., Pines, A. *Chem. Phys. Lett. 93*, 504 (1982).

276. Wimperis, S., Freeman, R. *J. Magn. Reson. 62*, 147 (1985).

277. Wong, T. C., Rutar, V. *J. Magn. Reson. 63*, 524 (1985).

278. Rutar, V., Wong, T. C. *J. Magn. Reson. 71*, 75 (1987); *63*, 524 (1983).

279. Rutar, V., Guo, W., Wong, T. C. *J. Magn. Reson. 69*, 100 (1986).

280. Rutar, V., Wong, T. C., Guo, W. *J. Magn. Reson. 64*, 8 (1985).

281. Shaw, D. *Fourier Transform NMR Spectroscopy*, Elsevier, Amsterdam, 1976.

282. Martin, M. L., Martin, G. J. Delpuech, J.-J., *Practical NMR Spectroscopy*, Heyden, London, 1980.

283. Chandrakumar, N., Nagayama, K. *J. Magn. Reson. 69*, 535 (1986).

284. Nagayama, K. *J. Magn. Reson. 69*, 508 (1986).

285. Patt, S. *Magn. Moments 2(1)*, 8 (1986).

286. Haupt, E. *J. Magn. Reson. 49*, 358 (1982).

287. Wesener, J. R., Günther, H. *J. Magn. Reson. 62*, 158 (1985).

288. Thomas, D. M., Bendall, R. M., Pegg, D. T., Doddrell, D. M., Field, J. *J. Magn. Reson. 42*, 298 (1981).

289. Pegg., D. T., Bendall, M. R., Doddrell, D. M. *J. Magn. Reson. 44*, 238 (1981).

290. Bax, A. *J. Magn. Reson. 52*, 76 (1983).

291. Bernassau, J. M. *J. Magn. Reson. 62*, 533 (1985).

292. Nielsen, N. C., Bildsøe, H. Jakobsen, H. J., Sørensen, O. W. *J. Magn. Reson. 66*, 456 (1986).

293. Sklenář, V., Nejezchleb, K. Starčuk, Z. *J. Magn. Reson. 69*, 144 (1986).

294. Bildsøe, J., Dønstrup, S., Jakobsen, H. J., Sørensen, O. W. *J. Magn. Reson. 53*, 154, (1983).

295. Jakobsen, H. J., Sørensen, O. W., Bildsøe, H. *J. Magn. Reson. 51*, 157 (1983).

296. Jakobsen, H. J., Bildsøe, H., Dønstrum, S., Sørensen, O. W. *J. Magn. Reson. 57*, 324 (1984).

297. Nielsen, N. C., Bildsøe, H., Jakobsen, H. J., Sørensen, O. W. *J. Am. Chem. Soc. 109*, 901 (1987).

INDEX

Numbers in **boldface** indicate pages where primary description of topic appears.
Numbers in *italic* indicate pages where topic is illustrated in figure.

Artifacts, **58**

BIRD, 63, 81, *83, 90,* 151, **183**
Boltzmann distribution, 154

Chemical shift correlated spectra, *see*
    COSY
Chemical shift resolved spectra, 30, 33, 50
CIDNP, *11,* 19
COCONOSY, 142
Coherence, 9, 154
  coherence echo, 179
  coherence transfer, **115**
COLOC, 89, *90–92,* 150–151
Correlated spectra, 24–25, *26,* **33,** *40,* 49,
    71, *75–76, 79, 82,* 87
COSY, *2, 11,* 71, **97,** *99–101, 103, 107, 113,*
    118, 121–122, 146, 148, 150–151, 198
COSY-NOESY, 142
CYCLOP, 106, 168, 198

Decoupling:
  during detection, *79, 82*
  during evolution period, 78, *82*
  during mixing period, 81
Delay, 13, 44, 77, *90*
  with continuous decoupling, 174
  without decoupling, 173
DEPT, 62, 81, 88, 96, 151, **180,** *182,* 206
Detection:
  anti-echo, *see* Detection, P-type
  echo, *see* Detection, N-type
  N-type, **189,** *197*
  P-type, **189,** *197,* 199
  quadrature, 42, **84,** 162, 168, 186, **189,**
    193
  single, 167, 186

Detection period, 10, 12, 14, **175**
Detection time, 10, 45
DOUBTFUL, 131
DQCOSY, *121*
Dwell time, 187

Evolution time, *10,* 11–14, 24–25, 41, 43
Exchange spectra, *11,* **132,** *137–138*
Exchange spectroscopy, *11*

Filtering, digital, **45,** *46*
Flip angle, 165, 203–204
FOCSY, 71, 108
Fourier transforms, **158,** *159*
Free induction decay (FID), 156, *157, 187*
FUCOUP, 74, 146, 151

Gated decoupler methods, 50

H,X-COSY, *72,* 85, 146, 148, 150–151, *203*
HETCOR, 85
Heteronuclear multiple quantum coherence
    (HMQC), 97
HOESY, 143, 151
HOHAHA, 108

Imaging, *11*
INADEQUATE, 72, 89, 120, **123,** *125,*
    127, 130, 146, 148, 150–151, *203*
INEPT, 62, 81, 87, 96, 151, **180,** *182,* 206
INSIPID, 130
Interferograms, 9, 15, 30, 42, 156

J cross-peaks, 139
Jeener, 4, 7, *8, 11,* 13, 25, 98, 135
J-resolved spectra, *5, 11,* 17, 24, *25,* 31, 50,
    69, **95,** 146, 148